P.A.S.S. C.A.L.F.

EIGHT BEHAVIORS OF SALES SUCCESS IN AN AGRICULTURAL DEALERSHIP

P.A.S.S. C.A.L.F.

EIGHT BEHAVIORS OF SALES SUCCESS IN AN AGRICULTURAL DEALERSHIP

by Frank Lee

©1997, Sales Academy, Inc., Flower Mound, Texas. ALL RIGHTS RESERVED. No part of this book or any of its contents or collateral materials may be reproduced by any means or for any reason whatsoever without prior written permission from the holder of the copyright. No part of the content may be reproduced in whole or in part, stored in any retrieval system now known or to be known, transmitted in any form or by any means now known or to be known, photocopied by mechanical, electronic, or any other means now known or to be known, electronically recorded, or reproduced using technology now known or to be known, for any reason whatsoever, without prior written permission from the copyright holder.

Sales Call Reluctance and Fear-Free Prospecting & Self-Promotion Workshop are registered trademarks, and Yielder is a trademark, all of Behavioral Sciences Research Press, Inc., Dallas, Texas, USA. ALL RIGHTS RESERVED. Used with permission.

ISBN # 0-9701399-0-X

Published by Sales Academy Press
2628 Timberhaven Drive
Flower Mound, Texas 75028
1-800-898-3743
www.sales-academy.com

First Published 1997
Reprinted 2000

SALES ACADEMY, INC.
HELPING GOOD SALESPEOPLE BECOME BETTER

ACKNOWLEDGEMENTS

Many people contributed to the writing of this book. Bob Honzik, former Division Manager and now retired, never stopped encouraging me. He provided many opportunities for me to further my knowledge of this wonderful industry.

My good friend Bob Hilleque has been an inspiration. He helped educate me in the ways of the agricultural industry, a chore at times. I suspect I was a challenge for him. If he could educate me, he can educate anybody.

There are so many people I should thank, but I would need several pages to include all their names. Rather, let me say one collective "thank you" to all the people who have helped me gather the data and experiences to be able to write this book.

There were many dealer salespeople, sales managers and owners whom I have met and worked with over the years. Each played an important part in getting this book written. This book is dedicated to helping them as they have helped me.

My wife, Pam, helped by accepting my heavy travel schedules and weeks on the road without complaining. She has stood by me and supported me throughout.

I am grateful to my two dogs, Baloo and Coco, for not biting me when I come home after many weeks away.

Finally, I must thank my two beautiful children for telling me something every father would be proud to hear. They said they thought I had an interesting business and wished they could be like me. Now that's something!

CONTENTS

FOREWORD

PREFACE   1

INTRODUCTION:
WHAT ARE SUCCESSFUL BEHAVIORS?   5
WHAT DOES DOCTOR LAURA KNOW
ABOUT SELLING?   10
ON MARKET SHARE, BOTTOM LINES AND
OTHERFANCIFUL STUFF   12

BEHAVIOR #1 – PLAN YOUR DAY, MAP YOUR
TRAVEL AREA   15
THE FINANCIAL IMPLICATIONS OF PLANNING   16
THE TROUBLE WITH PLANNING   17
A SIMPLE PLAN   18
THE LOST SALESPERSON   20
MAP YOUR TRAVEL AREA   23
WHAT IS THE BEHAVIOR?   24
WHY IS PLANNING A SUCCESSFUL SELLING
BEHAVIOR?   25
IF YOU ARE NOT AN AGRICULTURAL
SALESPERSON   25

BEHAVIOR #2 – APPOINTMENTS PAY ... 27
WHAT IS AN APPOINTMENT? ... 27
DEFINING AN APPOINTMENT ... 28
ON A RAINY DAY IN RURAL AMERICA ... 29
THE SPECIFIC PURPOSE OF THIS APPOINTMENT ... 30
DO FARMERS MAKE APPOINTMENTS? ... 30
WHY MAKE APPOINTMENTS? ... 31
WOULD YOU PREFER TO HAVE APPOINTMENTS? ... 31
DO FARMERS KEEP APPOINTMENTS, AND IF NOT, WHY NOT? ... 32
MAKE THE APPOINTMENT IMPORTANT ... 33
SOME ADVICE ON MAKING APPOINTMENTS ... 34
WHAT IS THE BEHAVIOR? ... 35
WHY IS MAKING SELLING APPOINTMENTS A SUCCESSFUL SELLING BEHAVIOR? ... 35

BEHAVIOR #3 – SOLVE PROBLEMS ... 37
WHY CUSTOMERS BUY ... 37
LOOK FOR PROBLEMS ... 40
$4 MILLION OF NOTHING ... 40
WHAT IS THE BEHAVIOR? ... 41
WHY IS SOLVING PROBLEMS A SUCCESSFUL SELLING BEHAVIOR? ... 41
IF YOU ARE NOT AN AGRICULTURAL SALESPERSON ... 42

BEHAVIOR #4 – SIX PROSPECTS ... 45
WHAT IS ASTRONAUT EQUIPMENT? ... 45
HOW EXPENSIVE IS ASTRONAUT EQUIPMENT? ... 45
WHY DO DEALERS ACCEPT ASTRONAUT EQUIPMENT? ... 47
HOW DEALERS COPE WITH ASTRONAUT EQUIPMENT ... 48
WHAT IS THE SOLUTION? ... 48
THAT *#*@* SALES MANAGER JUST COST ME ANOTHER DEAL! ... 50
OTHER POSSIBLE SOLUTIONS ... 51
SALESPEOPLE ARE BUSINESS PEOPLE, TOO ... 52
WHAT IS THE BEHAVIOR? ... 52
WHY IS SIX PROSPECTS A SUCCESSFUL SELLING BEHAVIOR? ... 52

IF YOU ARE NOT AN AGRICULTURAL SALESPERSON	53
SELL INTO THE FUTURE	54

BEHAVIOR #5 – COLD CALLS 55
WHAT IS A COLD CALL? WHAT IS A COMPETITIVE OWNER CALL?	55
COLD CALLING IS LIKE BUYING A LOTTERY TICKET!	57
WHY MAKE COLD CALLS?	57
HOW IS IT POSSIBLE?	58
A STUDY IN INDIFFERENCE	60
HOW DO YOU MAKE COLD CALLS?	60
ONE-TRIAL OR ZERO-TRIAL LEARNING?	61
WHAT IS THE BEHAVIOR?	62
WHY IS COLD CALLING A SUCCESSFUL SELLING BEHAVIOR?	62
IF YOU ARE NOT AN AGRICULTURAL SALESPERSON	63

BEHAVIOR #6 – ASK WHY 65
FOUR LEVELS OF SELLING	65
QUESTIONS IN SELLING	67
WHAT DO "ASK WHY" QUESTIONS DO?	67
EXAMPLES	68
DON'T STOP WITH WHY	69
WHY ASK WHY BEFORE THE SALE?	70
DEVELOP A DYNAMITE SALES PRESENTATION	71
WHY ASK WHY AFTER THE SALE HAS BEEN MADE?	71
THE GOOD SALESPERSON	72
WHAT IS THE BEHAVIOR?	73
THE KEY TO ASKING QUESTIONS	73
WHY IS "ASK WHY" A SUCCESSFUL SELLING BEHAVIOR?	74
IF YOU ARE NOT AN AGRICULTURAL SALESPERSON	74

BEHAVIOR #7 – LEARN ABOUT COMPETITION 77
HOW MUCH DO YOU KNOW ABOUT YOUR COMPETITION?	77
WHY SHOULD YOU LEARN ABOUT YOUR COMPETITION?	78
HOW DO YOU LEARN ABOUT YOUR COMPETITION?	78

WHAT COLOR IS YOUR PROFILE? 80
HOW DIFFERENT ARE YOU FROM YOUR
COMPETITORS? 81
ELIMINATING COMPETITION 82
WHAT IS THE BEHAVIOR? 84
WHY IS LEARNING ABOUT YOUR COMPETITION A
SUCCESSFUL SELLING BEHAVIOR? 84
IF YOU ARE NOT AN AGRICULTURAL
SALESPERSON 85

BEHAVIOR #8 – FIRST-TIME TRADE 87
DOING THE IMPOSSIBLE 87
WHAT HAPPENS IF YOU DON'T CLOSE THE
FIRST TIME? 88
HOW? 89
DEALING WITH PRICE OBJECTIONS 89
WHAT IS THE BEHAVIOR? 92
ARE YOU A YIELDER ? 92
WHY IS FIRST-TIME TRADE A SUCCESSFUL
SELLING BEHAVIOR? 93
IF YOU ARE NOT AN AGRICULTURAL
SALESPERSON 93

CONCLUSION 95

FOREWORD

Who needs another sales training program?

There are already more than 2000 sales training books, cassettes, seminars, programs and workshops, all competing for the same dollars: *yours*. Most contain nothing new, despite bold claims to the contrary.

But Frank Lee's book fills an important need. It departs from the ultra-popular, broadbrush, generic approaches to sales training, which are rightly criticized for touting extreme results whether you sell insulation or insurance, clothing or carpets, dinner rolls or Rolls Royces. Frank knows that sales *in genera* won't sell, so he has written a specific book, with specific information, to help salespeople in the agricultural equipment market.

Frank is not merely another author writing *about* selling. He passionately loves to sell, and he lives what he teaches every day. More importantly, against the odds, despite discouragement and put-downs from Australia to Atlanta, Frank has proven where it counts, with bankable results, that he is a master seller. And he does it with poise and style.

So who needs another book on selling? It doesn't matter if you are a wet-behind-the-ears neophyte or a world-weary veteran. If you are in farm equipment sales, if you have a teachable heart and the will to sell more, chances are you do.

George W. Dudley
Co-author, *Earning What You're Worth?*
Dallas, Texas
1997

ANOTHER PERSPECTIVE

I met Frank Lee as he was presenting his Fear-Free Prospecting Program sponsored by his company, Sales Academy. Fear-Free Prospecting was a successful program and was the first step in leading and training retail sales people away from "running the trap line" - calling on the same John Deere customers year after year and using the "good old boy" system for closing a sale that usually impacts profitability.

Frank Lee, on his own, chose to fully acquaint himself with our industry, particularly at the retail level. He spent considerable time with John Deere retail sales people calling on farmers. He observed, first hand, the good, the poor and the indifferent to retail sales. Frank and myself, together with other John Deere employees, had ongoing discussions and studied approaches identifying our retail sales problems and developing a new approach. Frank Lee, with his worldwide marketing experience, was the first to see we were in desperate need of behavioral changes at the dealer level. The rapidly growing level of sophistication among farmers and equipment was leading to a widening knowledge gap between the retail sales person and today's professional farmer. The result of Frank's work became the "Successful Dealer Behaviors Program."

This program was tried at three John Deere dealerships. Each had unique marketing and personnel challenges. Owners, department managers and employees fully participated in the new program. All found this was not a laborious or time-consuming effort. Results were almost immediate and beneficial to employers, employees, their customers and John Deere Company.

Results from the "Successful Dealer Behaviors Program" united dealer employees into making a common effort to develop a broader customer base while acquiring a clear understanding of how each employee contributes to the financial health of a dealership. Backbiting between departments and employees ceased. This improved department and customer relationships. All in all, the three pilot dealerships exceeded their prior expectations in sales and financial performance, due solely to successful behavioral changes. Results from the pilot program were obvious. We offered this program to additional dealers. Results mirrored the three pilot programs. We also saw successful behavioral changes have a personal effect on many of the participants... not just financial improvement but also in life-styles and relationships.

Frank Lee gives each dealer and each employee of the dealership his personal attention. Those who participate in the "Successful Dealer Behaviors Program" will benefit from Frank's keen knowledge and ability to lead people in acquiring personal growth.

Bob Honzik
Division Manager (Now Retired)
John Deere Company
Minneapolis

PREFACE

My first experience with the agricultural industry was on my maternal grandmother's farm in South Africa, near Oudtshoorn in the Karoo. I still remember seeing her John Deere tractor when I was a kid visiting her. Back then I was not allowed to actually drive it. That was more than 40 years ago.

Years later, one of the first shocks I received when I traveled with agricultural equipment salespeople was the discovery that the farming population had changed. I expected to see people like my grandmother on the farms we called on. What I found instead were young business people. Computer savvy and highly educated, they knew more about the world of business than my grandmother ever knew or wanted to know. There were still the good old boys, to be sure. I'll never forget the seventy-year-old farmer who shook my hand—and nearly broke it. But he was part of a disappearing breed.

Another shock came when I realized just how much farm equipment had changed over the years. Today there are machines that communicate with satellites and can tell the farmer where to plant. The Atlanta branch of John Deere allowed me to sit inside a new combine that stood in their training center. Again, I was not allowed to drive. (Some things never change!) It felt like a scene from *Star Wars*. Sitting in the "cockpit"—for that's what it felt like—gave me a tremendous sense of power. It felt almost as if this mighty machine were made for some futuristic farm in the movies. Grandmother would have loved this! She would not have understood it, but boy, would it have looked good on her farm in the dusty Karoo!

The sophistication of the new equipment will, I believe, drive the industry in the years to come. Already it has taken a toll. Many medium-sized farms are now consolidating. On a radio news program in North Dakota I recently heard that the number of large and small farms is increasing, while the number of medium-sized farms is decreasing. Dealers I speak to agree that the number of customers available to them is shrinking. In Wisconsin, dairy farms are closing at the rate of four per day. They are not necessarily going out of business. The corporate farmer is buying them up, or they are consolidating. Two things, therefore, have to concern the agricultural salesperson—a rapidly growing level of sophistication (farmers and equipment) and a rapidly shrinking customer base.

Agricultural sales is no longer business as usual. The dealer salesperson has to change. He must rise to the market's level of sophistication or the market will preclude him. The "good old boy" selling style may be on its way out.

Some savvy salespeople in dealerships have realized this and have already taken steps to ensure they will be around in the long term. These are the top sellers who adapted when they had to and who will lead the charge into the next millennium. Unfortunately, most dealer salespeople live in a state of denial. They cannot or will not see the impact of the changing market on their own careers. While they pay lip service to change, they really have done nothing to improve their own chances. This is unfortunate, because the time is rapidly approaching when they will be overtaken by the rising tide.

Ironically, right now business could not be better for many of them. They are increasing sales in record numbers. Market share is improving. What they tend to forget, however, is that business is good for *everyone* right now. I have met only two dealers in recent memory who told me business was bad. Both had internal problems that had nothing to do with the economy. For the rest, life is good.

Yet even as dealers crow about the boom years, used inventory builds up on their lots. Profits are not as high as they should be. Some work on ridiculously low margins. Many depend on their service and parts departments to make up for low margins on whole goods sales.

In a way, dealers are very much like the farmers they serve. They worry about the future. Beneath the shiny happy surface, there is a sense of unease in many dealerships. The boom market has been going on too long for their comfort. I hear constant reminders about the cyclical nature of the business. Dealers bring up the bust years of the Eighties more and more often. They seem to be expecting a market decline, although nobody is prepared to predict when it will come.

The clever dealer salespeople will not wait to see which way the market goes. They will arm themselves with potent weapons to fight any coming war. The really smart ones have already taken action. Some dealers have started to recruit and train young tigers to sell. A dealer told me he had interviewed one such salesperson. This tiger was already working for one of his competitors. His words to me afterwards were: "I have seen the future and it scares me. My salespeople could not compete with him."

There are many financial and market reasons salespeople in any industry should improve their selling abilities. Two of the most basic are that customers expect it, and it is good business. Customers have a right to expect salespeople to be up to date and to be able to rely on them for expert advice. It is the job of the salesperson to provide this expertise and to help customers reach better buying decisions. It is good business to be the best salesperson in your area.

Yet even as salespeople and their managers talk about the need to improve, there seems to be frustration over how to go about it. They have attended sales courses. They have spent time at their industry conventions. They have had buzzwords like *profiles*, *competitive owners* and *customer satisfaction* drummed into them for years. They are not stupid. They know they should be doing many things. They simply do not know how.

If this describes the state of your selling career, this book is for you. It's part of the "how." Perhaps, instead of talking about change or cerebralizing, it is time to actually practice some successful behaviors that will move you closer to your goals. To all the salespeople in agricultural dealerships (and others) who would like a practical way to improve their sales and earnings, I hope you can find it in the pages of this book.

INTRODUCTION:
WHAT ARE SUCCESSFUL BEHAVIORS?

"I'm the best salesperson you've ever seen. Heck, you've *never* seen a salesperson like me before. I'll make you rich and famous. And I'm coming to work for your dealership tomorrow.

"I already think I'm a success; but what would I have to do so *you* would consider me a successful salesperson?"

I looked around the room for responses. The ten agricultural sales managers grinned. They had all been here before. They knew they were going to make me squirm. Tim, a red-faced, heavyset, jovial type, shot back without hesitation: "Sell!"

"I can do that," I said smugly. "I told you I was the best. I'll go out first thing tomorrow and sell a tractor. I have this relative who farms close by. He's not really a relative. He just thinks he is because we look alike. I'll sell him a tractor in a heartbeat. Now am I successful?"

"What else are you going to do?" Tim asked.

"What else do you want me to do?"

P.A.S.S. C.A.L.F.

"You're going to have to make more than one sale. Besides, you'll have to sell at a good margin."

"I can do that, too," I replied. "This relative really likes me and he'll pay whatever I ask. So now I have your margin. Am I successful?"

"You also have to create good customer satisfaction," said Tim. "Your relative has to have a good relationship with you."

"I can do that," I told him. "I'm very good at creating relationships. I'll take him for a beer after the sale and we'll have a good time together. He'll even invite me home for dinner. Now am I successful?"

"You also have to make more than one sale," he retorted. "You're not successful until you make many more like that."

"How many relatives do you think I have?" I asked. "I just made one very good sale."

John, the manager from North Dakota, piped up: "You also have to accomplish your goals."

"I just did," I told him. "I made one sale. That was my goal for the year. Now am I successful?"

"What about the goals of the dealership?" asked Dave. "You need to accomplish those, too."

"Those are your goals, not mine."

"Well, we have some goals for you to accomplish, and you have to reach them, too."

"Like what?"

"You'll have a target. You should also sell additional things to your customers." This was Jim.

"I can do that," I said proudly. "I'll sell my relative some toys. And before you tell me about margins again, he'll pay good margin. Now am I successful?"

"No." Rick, who had been quiet all this time, was starting to get riled. "There's more to being a success than that. You also have to make sure all the paperwork is in order."

"Oh, that. Well, okay, if you insist. What else do you want me to do?"

"You must continue selling," said Jim. "Lots and lots more."

"Well, now, let's set this to rest. Exactly how much do you want me to sell?"

Jack, sitting in the corner of the room, jumped in. "I want you to sell a million dollars."

"Done!" I told him. "I'll get all my relatives together tomorrow and make a million dollar sale to all of them. Can I now take the rest of the year off?"

"No!" Rick was really getting upset with me now. "That's tomorrow. You also need to do it the day after and the day after that."

"Sorry, Rick," I told him. "You want too much from me. I'm going to go to work for Tim. All he wants is for me to make one sale."

"Oh, no," Tim jumped in quickly. "I want you to sell more than that, too. Besides, I want you to have high goals. I want you to meet your goals. I want you to be honest in your dealings with your customers. I want you to make sure all your customers are properly taken care of. I want you to make them happy."

"You want a saint, not a salesperson," I interrupted him. "I thought all you wanted from me was to make that one sale. Perhaps I won't work for you after all. Jim, can I come to work for you?"

Jim gave me a sly grin. "Yes. But I want the same as Tim. In addition, I want you to make good profits on your deals."

By now everyone in the room was putting in his two cents. One by one, they added to the list of things I needed to do in order for them to consider me a successful salesperson. The list kept growing and growing. I looked at them in bewilderment. They knew they had me. I was not going to get away with me-

7

P.A.S.S. C.A.L.F.

diocre performance in their dealerships. Eventually, I stopped them.

"Do you know what the problem is?" I asked. "The problem is not with me. The problem is not with you. The problem is this word 'successful.' It's not even a proper business term. It is too open to interpretation. As you have seen, we all have different views of what constitutes selling success. We could go on for days and keep adding to the list. We started out simply enough with one definition—just sell. Then we developed it to the point where it would take a superman to satisfy all of you."

The room had gone quiet. They were wondering where I was going.

"The only time this word makes any sense is when we start to talk in terms of 'successful behaviors.' Let's change the scenario a bit," I suggested. "Forget that hotshot salesperson. He doesn't really exist. He never did. Instead, here I am. I'm not a superstar. I'm just an honest-to-God hardworking agricultural salesperson. I don't have any rich relatives. But here's what I will do in your dealership, starting tomorrow.

"From day one, I will call at least five of your existing customers each day for no other reason but to ask them if they would like to buy something more from your dealership.

"From day one, I will make at least one appointment a day with someone I have a chance of selling something to that day.

"Each day I will make at least two cold calls on people who have never been in your dealership before to tell them we exist and ask if they would like to come in and find out what we have to offer.

"Each day I will make a point of contacting at least two competitive owners to let them know we have equipment as good as, if not better than, what they already have and to invite them to examine their options.

"Each week I will collect at least two profiles because I know the value of profiles. I know I can use them to create future business.

"Each week I will do at least one on-farm demo.

"And, as I do all these things, I will ask everyone I contact for the names of others they can refer me to.

"Can you see where I'm going with this?"

They knew. The room had quieted. They were nodding their heads.

"If I do all these things, and I do them consistently, every day, will I sell?"

"Yes."

"Sure."

"No doubt about it."

"Is there any doubt that I will sell?"

They shook their heads.

"Do I have to be the most polished salesperson in the world?"

Again they shook their heads.

"So what you're saying is that all I have to do is practice these behaviors and do them on a consistent daily basis and I will sell?"

They nodded agreement.

"No doubt about it," said Tim. "Matter of fact, you do things like that every day, you've got a job for life with me."

"Not only will I sell," I said, driving my point home, "I will outsell the smartest salesperson who knows ten times more than I do but who sits on his butt in the office waiting for the phone to ring. I will outsell the cleverest salesperson who talks a good game but does none of these things. You've already got some of those. They spend their time driving around the countryside aimlessly, without plan, going where the road takes them.

P.A.S.S. C.A.L.F.

WHAT DOES DOCTOR LAURA KNOW ABOUT SELLING?

I'm one of those channel surfers. In my case, I also surf the channels on my car radio. Sometimes a piece of music will grab me and I'll settle there for awhile. Then another song comes on that I may not like as much and I start surfing again. Occasionally, I'll land on a talk show and, depending on the amount of animation I hear, I'll either move on or stay and listen awhile. Not a big fan of talk shows, I usually don't stay very long. This was how I happened on Dr. Laura. She is one of those psychological types who has answers for everything. I had caught parts of her show before and liked her sense of humor even when I didn't agree with the advice she gave. At least she was entertaining. One time, however, I was struck by her advice.

A lady had called in to complain that the romance had gone out of her marriage and her husband was no longer having sex with her. Dr. Laura quickly got her to admit that she loved her husband and wanted to have romantic interludes with him. She wanted to have sex and she wanted to enjoy it. However, the situation had deteriorated to the point where they hardly even spoke to each other. Dr. Laura asked her, "When are you going to have what you want?" The caller didn't know. She rambled a bit about thinking about this for a long time and not finding any solutions. She could not get back the feelings she desperately wanted, the feelings they had had before. Dr. Laura's next remark took the caller (and me) by surprise.

"If you want what you say you want, you are not going to get it by thinking about it. You will get it by starting to *act* romantically toward your husband. Talk to him as if you love him. Touch him when you talk to him. *Act as if you're in love.* And guess what? The feelings will follow. It will not work the other way around. Thinking about it won't make it happen. Act first, and the feelings will follow."

I don't know much about Dr. Laura, but I suspect she knows a thing or two about behaviors. Thinking about success won't make it happen. Practicing the right behaviors will bring the feelings. Don't trust me. Trust Dr. Laura.

"We're not talking about rocket science. We're talking about very simple behaviors, aren't we?

"If it's that simple, how come the salespeople in Iowa are not doing it? How come the salespeople in the dealership down the road from you are not doing it? How come your salespeople are not doing it? And how come you're not doing anything about it?"

I often use this exercise to teach managers the importance of successful behaviors. It is an adaptation of a technique taught to me by the master of behaviors, George W. Dudley, a world-renowned behavioral scientist who developed the TriPart Model to demonstrate the importance of the right behaviors. George tells me he developed this model to show the relativity of testing in the hiring process. It was by studying under George and learning the proper use of this model that I was able to develop this exercise.

Often sales managers and sales trainers spend a great deal of time "educating" salespeople. Their salespeople learn all the right words. They learn all the right closes. They become extremely clever scholars. But they still do not succeed. Additionally, they are exhorted to develop incredible virtues. The gurus of the sales world teach us to strive toward character traits they themselves may possess in limited quantities. They teach us that we will not succeed unless we possess these virtues.

I recently read a book by a well-known motivator. Her subject was the 14 exciting secrets of success. When I was through, I felt the same as I have often felt after reading these so-called motivational books: there was nothing new, different or exciting here. This author told salespeople that to be a success they needed to be motivated, goal-driven, and ethical. They needed to build trust, make the customer feel important, and develop rapport.

P.A.S.S. C.A.L.F.

They should give if they wished to receive. Nothing we have not heard before.

These are the same principles our parents taught us. There is nothing wrong with them. They are good and worthwhile and apply to other professions besides sales. However, sales motivation books tend to imply that you will fail in sales if you do not reach these very high moral levels. If only we could!

What about us mere mortals? Do we have no chance to succeed in sales? What if we are not as super-motivated or as highly goal-driven as the experts would have us be?

This book is not for super-achieving saints. It is for us mortals who have to make a living at sales while we are trying to put our spiritual lives in order. It's based on the belief that even unsophisticated, unpolished salespeople can outperform the most polished, informed and educated salesperson if they practice the right behaviors on a consistent daily basis.

Even though I conceived this book for agricultural salespeople, all the behaviors I recommend apply to other sales situations. There is a special note at the end of each chapter showing their relevance to other kinds of sales.

ON MARKET SHARE, BOTTOM LINES AND OTHER FANCIFUL STUFF

"Market Share" is a complex term understood by marketing gurus and thrust upon unwary salespeople and sales managers. Most sales managers and their salespeople are not concerned with market share. They care about making one sale at a time. They know they cannot make the second sale until they make the first.

"Bottom Line" is an aggressive term bandied about by aggressive finance managers. It sounds good. Salespeople in the trenches neither understand nor care about the bottom line of their company. They are more concerned about having to handle that rude prospect one more time.

Even if they did care, individually they cannot make any significant difference to either market share or the bottom line of the company. Theirs is a micro view of the entire process. So why do so many sales managers think they are motivating their salespeople when they throw these misunderstood intangibles at them?

They would do better to concentrate on simple, successful behaviors their salespeople could apply on a consistent daily basis. If they could encourage these behaviors in their salespeople and make them habits, they would do far more to motivate them. They should also make these behaviors conditions of employment for all new hires.

Once these behaviors become habits, once they become an integral part of the way they do business, then the company and everyone in it can reap the benefits of a terrific market share and a booming bottom line.

In this book I take a micro look at the sales process. I'm not concerned with everything that happens in a company. I'm not even concerned with the entire sales process. I assume your technical knowledge and basic sales skills. This book is not even an advanced sales course. It is far more basic than that. It is simply good business.

In these pages I'll concentrate on a single premise: If you practice the right sales behaviors and make them habits, you will succeed in sales. No matter what you sell. No ifs, ands or buts. Most selling problems can be resolved by applying these behaviors.

The eight most important successful selling behaviors are covered in this book. There are, of course, many more, and I hope the reader will discover them and put them into practice. However, if you are an agricultural salesperson and you put these

P.A.S.S. C.A.L.F.

eight behaviors into practice, you will succeed in sales beyond your wildest dreams.

This book was written especially for agricultural implement salespeople in a dealership. These are the people I have worked with for the last four years. They taught me what I needed to know about selling farm equipment. P.A.S.S. C.A.L.F. started out as a workshop that was part of the four-month Successful Dealer Behaviors Program that my company, Sales Academy, conducts for select dealers throughout the United States. Today there are agricultural salespeople out there already practicing these behaviors—and killing their competitors.

If you are not in agricultural sales, don't despair. These behaviors can be adapted to whatever it is you sell. They can turn your career around, too. The acronym for these behaviors is P.A.S.S. C.A.L.F. It is purely agricultural. For those willing to give them a try, they could lead you to the fatted calf.

BEHAVIOR #1

PLAN YOUR DAY, MAP YOUR TRAVEL AREA

Planning simply means knowing ahead of time the things you intend to do each day. It is a roadmap that leads you in a direct path to your goals. Your roadmap reminds you where you have to be and when. *Planning is the basis for all successful behaviors.*

Sometimes, when I talk about planning, salespeople tell me I don't understand how busy they are. The most common excuse I hear is this: "As soon as the phone rings, my plan for the day goes down the toilet. I spend my day putting out fires." In other words, "I'm too busy to plan." Yet planning is ideal for busy people. It's not for those who have plenty of time. If you're not busy, what have you got to plan? Planning helps busy people get on track and stay on track.

P.A.S.S. C.A.L.F.

THE FINANCIAL IMPLICATIONS OF PLANNING

Why should agricultural salespeople plan their days? Because drive time is expensive! Many salespeople tell me they spend up to 50% of their time in their pickups, going from one farmer to another. That number may be a bit high. Part of that 50% may not actually be in the pickup but used up by other non-productive work (time wasting).

How expensive is this time? Let's see.

If:
You currently earn $40,000 per year;
and you work 11 months of the year, 25 days of each month, 10 hours per day;

(Face it: nobody works every single day of the year. There are weekends, vacations, sick leave, public holidays and birthdays. There are also times when agricultural salespeople work very long hours, and they are quick to point this out. My numbers will still prevail if we average it out over the year.)

Then:
$40,000 divided by 11 = $3636 per month
$3636 divided by 25 = $145 per day; $145 divided by 10 = $14.50 per hour

If you are like the typical agricultural salesperson, you spend 5 hours per day in your pickup (or wasting time). This means your driving time costs you 5 x $14.50 = $72.50 per day or $20,000 per year. Would you pay someone $72.50 every day just to drive around?

Let's look at this a little differently. The other 50% of your time actually produces the sales that earn you the $40,000 per year. Therefore, your selling time is not worth $14.50 per hour. It's worth twice that, or $29 per hour. This means if you spend all your time selling and no time in your pickup, your earning capacity has doubled to $80,000 per year.

Of course, that isn't possible. There will always be drive time. There will always be some unproductive time. You will always waste some time.

But let's see if we can reduce your unproductive time. Let's save just half of the unproductive time—or 2 hours per day—through proper planning. This will give you an extra 2 hours per day of selling time.

2 hours @ $29 per hour = $58 per day extra.
$58 x 25 days = $1450 per month extra.
$1450 x 11 months = $15,950 per year extra.

By spending just five minutes each day plus 30 minutes each week on a planning exercise, you open the door to an additional $15,950 per year of potential income. This means you are actually being paid $5.92 per minute to plan. That's $355.20 per hour. Is it worth it?

When you include savings on gas, wear and tear on the pickup, and other driving expenses, you can see why even your dealer principal will applaud you.

THE TROUBLE WITH PLANNING

One of the biggest problems with planning is that salespeople often try to over-plan. They attempt to plan every minute of the day ahead of time. That type of planning is doomed from the

P.A.S.S. C.A.L.F.

start. It does not take into account the fires that spring up each day that have to be put out. It does not leave any elastic time.

Some companies sell planners that are so complicated it takes two days of classroom study to understand their process. Then it takes years to get it right. This is not to say their planning systems are not useful. They can be extremely useful, and for the person taking the time to learn the system they can help a great deal. However, I have met many salespeople who tell me they still don't understand the process after years of using it. Isn't there a simpler way?

A SIMPLE PLAN

What follows is a simple planning system that I have used for years. It has kept me on track through extremely busy days and has allowed me to get more things done than most people. It takes me about thirty minutes each Sunday to plan my week and five minutes at the end of each day to make adjustments for the next day. The time spent planning has helped create time to get things done and has paid off handsomely.

Start with a yellow legal pad. Use one page for each day of your workweek. Write the day at the top of each page. Divide each page into four segments: Phone Calls (these are the calls you intend to make that day), Calling Me (these are the people who said they would call you that day), Appointments, and To-Do's. You should have a page that looks something like this:

Eight Behaviors of Successful Selling

Monday, September 15, 1997

Phone Calls	Appointments
1.	8 am
2.	9 am
3.	10 am
4.	11 am
5.	12 pm
6.	1 pm
7.	2 pm
8.	3 pm
9.	4 pm
10	5 pm
Calling Me	To Do
1.	1.
2.	2.
3.	3.
4.	4.
5.	5.
6.	6.
7.	7.
8.	8.
9.	9.
10.	10.

P.A.S.S. C.A.L.F.

Now fill in the activities you have planned for each day of the week.

> ### THE LOST SALESPERSON
>
> It was one of those cool fall days when it felt better to be inside a pickup than outside. I had no idea where I was. A dealership had kindly allowed me to spend a few days with their salespeople so I could observe firsthand what they did and how they did it. This part of the country was new to me, and there were no landmarks to help me get my bearings. Joe, the salesperson, could have been taking me back to Dallas for all I knew. We had left the dealership around 8 a.m. It was now nearly 10 a.m. and still we had not stopped anywhere. I figured it was a long way to his first call. Then I realized something.
>
> There was one landmark I recognized. It was the highway I had come in on the night before. The reason this suddenly struck me was that we had just crossed it for the third time. We did eventually make some stops. We also crossed that same highway another three times before lunch. This told me that Joe had no idea where he was going. He was simply driving around. I could see he was thinking as he drove. Perhaps it was the fact that I was in his pickup with him that caused him to lose his way. I like to think it was because he was trying really hard to find "good" customers to call on so he could help in my education. I felt as lost as he.

Phone Calls – Fill in the names and numbers of people you intend to call each day. You could be making calls during drive time or at other times in the dealership. Or you could set aside an hour each morning to make as many of the calls as you can and use drive time to catch the others.

Calling Me – Write in the names of people who have said they would call you on the day they said they would call. They didn't say which day they would call? Get into the habit of asking customers or prospects when they will call you. Whenever someone says they will call you for any reason, always ask when. Then write their names on the appropriate day. If they don't call you on the day they said they would, put their names on the Phone Calls section of the next day. *Then call them.*

Appointments – This is where you fill in the names of the people you have appointments with each day. You could also fill in names of people in the area that you can call on if you have the time. (See Map Your Travel Area, later.) You should have at least one solid appointment each day (more on this in the next chapter), and the total number of appointments will depend on how busy you are. Be realistic about the number of quality appointments you can reasonably handle in a day. Keep in mind distances and travel time as you book your appointments.

To-Do's – These are the things you want to do each day. Even personal tasks should be included here, especially if they will be done during business hours. While sales managers frown on doing personal chores during working hours (and I agree with them), realistically speaking, some things simply must be done during the day. You're going to do them anyway, and you should at least have them in your planning schedule.

As you go through the day, draw a line through the items you accomplish. At any time you can see where you are in your planned day. If you get really busy, you have a plan to come back to.

At the end of each day, make adjustments to your plan. Transfer any notes you may have made on each page to appropri-

P.A.S.S. C.A.L.F.

ate places, such as your computer. Don't forget to transfer any tasks not accomplished to another day. Once you get the hang of it, this should take less than five minutes at the end of each day.

You should also have a monthly calendar to write in scheduled appointments made more than a week ahead.

That's it!

The value of this system is that you can plan an entire week at a time. If you see you have too many tasks for a particular day, you can spread the load over other days when you have less to do. For example, if you have six appointments on Tuesday, you may want to shift some phone calls or to-do's to other days. If you know you have customers or prospects calling you on Wednesday and you will be out of the dealership, you can alert your office personnel to have them call your mobile phone.

This simple planning tool is not only useful for you. If you give a copy of each day's lists to your office personnel, they can help you stay in touch with your customers. They will not have to guess where you are. All they have to do is look up your planning schedule. A customer needs to talk to you? Thanks to your plan, your office knows you will be in his area. Your office personnel can tell the customer and set a tentative appointment that you can confirm while in the area.

This simple planning system has many benefits.

1. It helps you see where you are and where you will be for one week at a time.
2. It keeps you on track even when you get so busy you don't know which way is up.
3. It makes sure nothing falls through the cracks. It alerts you when prospects who said they would call, don't call. It reminds you to call when you said you would.
4. It helps your office personnel stay in touch with you.

5. It makes sense of your busy day.
6. After a few weeks, it will tell you which days tend to be your most productive, which will help you plan even better.

MAP YOUR TRAVEL AREA

Most customers live many miles from your dealership and it takes quite a bit of time to get to their farms. This is why drive time cannot be eliminated altogether. However, it can be used more effectively if it is planned. Here is one simple, effective piece of advice: Always travel in a circle.

Start with a map of your travel area. This map should list every one of your customers and potential customers. It should be large enough to read easily. You may want to have two copies: one on the wall in your office and a smaller version in your pickup.

When you set appointments, keep this map in mind. It will help you avoid crisscrossing your territory. If you have an appointment with a farmer in one area of your map, schedule other appointments in the same area. Take into account the time you will need to get from one appointment to the next. You may find you have time to call on other prospects in the same area while going from one place to another.

Once you are in your pickup, you may as well see as many people as possible. You will waste a great deal of time if you make one call, go back to your dealership, make another call, go back to your dealership, and so on. This is what I mean by always traveling in a circle. After each appointment, look at the map to determine whom else you can call on in the same area. Look at the prospects you could be visiting on your way to your appointment and on your way back. You may be surprised at

P.A.S.S. C.A.L.F.

how many people you can actually see in a day. You can even call on those people you habitually drive by. If you're going to drive at all, make use of your drive time.

This sounds simple, and it is. However, I have driven with salespeople who seemed to tool around without any purpose, often revisiting places they had already been to. With proper planning, you should never have to visit the same place twice in a day.

Some "do's" and "don'ts":

1. Do know where you're going.
2. Do always drive in a circle.
3. Do look at your map before you start out and consciously plan to visit at least one other farmer in the area where your appointment is.
4. Do use your mobile phone while driving. (Observe safety on the road, of course.)
5. Don't visit the same place twice in one day.
6. Don't ever make just one call when you drive.
7. Don't crisscross your territory.

Mapping your travel area and planning your travel will save time and bring you more business.

WHAT IS THE BEHAVIOR?

Plan your day. Map your travel area.

WHY IS PLANNING A SUCCESSFUL SELLING BEHAVIOR?

Planning is for busy people. It's the basis of all successful behaviors. Planning allows you to take positive control of your day. It requires you to do something each day, each week. Eventually, it becomes a habit. When it does, it becomes the way you do business.

IF YOU ARE NOT AN AGRICULTURAL SALESPERSON

This chapter is not limited to agricultural salespeople. It has universal applications. Look at your business. Can this type of planning help you? Don't let its simplicity fool you. Most successful people use some type of planning. Some use more sophisticated planning systems. Some simply write little notes to themselves. However they do it, they do it consistently and with forethought.

There are many salespeople who give up on planning because of the time and effort it takes to maintain their plans. Perhaps you are one of them. Perhaps the answer is that your plan is too complicated. The solution may be a simple planning system such as the one described in this chapter.

One salesperson I know has described this system as "elegant in its simplicity" and has used it to double his income. Adapt it to your product and your career. You may be pleasantly surprised.

BEHAVIOR #2
APPOINTMENTS PAY

WHAT IS AN APPOINTMENT?

In workshops around the country, I have asked agricultural dealer salespeople about appointments. This is what they have told me.

How do you define an appointment?
Salespeople usually describe it as a meeting at a specific time, with a specific person, for a specific reason, at a specific place.

What are the reasons for the appointment?
There are many reasons. There are, for example, service appointments to make sure the customer has received what he has bought. There are informal appointments to build goodwill. There are selling appointments, demonstration appointments, collection appointments, and so on.

27

P.A.S.S. C.A.L.F.

Do farmers make appointments?
The responses are mixed. Some say no. They say the farmer is loath to make an appointment. Others say yes, the farmer prefers to make an appointment.

Why make appointments at all?
It saves time, gives you a definite purpose when calling and prepares the farmer for your call.

Would you prefer to have appointments?
Except for a few hardened old-school types, agricultural salespeople have no hesitation in saying yes.

Do farmers always keep appointments?
No.

Why not?
Responses vary on this one. Most make excuses for the farmer. I suspect they are talking from experience and frustration.

In this chapter, I will attempt to answer these questions from Sales Academy's point of view. In doing so, perhaps I can put appointments in a new light for you and demonstrate why making appointments is a successful behavior. Let's take the questions in order.

DEFINING AN APPOINTMENT

For the purposes of this behavior, I will limit myself to one simple definition of an appointment – a planned meeting with a specific person, at a specific time, in a specific place, for a very specific purpose.

ON A RAINY DAY IN RURAL AMERICA

It was one of those days when my most favorite of all places would have been in front of my fireplace. Instead, there I was, standing ankle deep in mud, soaked by the steady drizzle that fell gently on me. I was not even dressed for the occasion. Neither was my salesperson guide for the day, but at least he had boots on. Jeff looked equally miserable as he tried to make a sales presentation in the rain. The farmer seemed to be the only one enjoying this episode. He was knee deep in muck, joyfully shoveling soil from a trench while his helper operated a mechanical shovel close by. He would occasionally look at us and say just enough to get Jeff's hopes up. Then he would return to his task. He was playing games with us.

I felt sorry for Jeff. He was a young, ambitious salesperson with plenty of promise. He thought he would get a sale when all he got was ridicule. He eventually gave up just one second before I was going to drag him away. The farmer cheerily waved to us as we struggled up the embankment to where we had parked.

Jeff looked at me as we sat shivering in his pickup. "What could I have done differently?" he asked. I was impressed that he had already figured out that something had not gone right and that he had the humility to ask for advice.

"Did you make an appointment?" I asked him. He nodded yes.

"Perhaps you should have made it more important. It sure didn't seem important to that farmer. Next time you make an appointment with anyone, use the words 'Let's sit down when I get there.' It may help."

Jeff was clever enough to use this advice. We spoke at length on the way back. He was like a sponge. I was supposed to be observing, and here he was making use of me to increase his knowledge. He eventually got on the management team in record time. I suspect I will hear more about this highly intelligent young man who learned a valuable lesson in the rain.

P.A.S.S. C.A.L.F.

THE SPECIFIC PURPOSE OF THIS APPOINTMENT

I understand there can be many valid reasons for an appointment. For the purposes of this behavior, however, I will limit the purpose of the appointment to a sales appointment. You will soon see why this definition is crucial to the behavior.

In other words, this appointment will be with a specific person, at a specific time, in a specific place, and this person must be *someone I have a chance of selling something to today*. This is a selling appointment. Both my prospect and I should know this ahead of time. This is not a visit, a chat call, or a drop-in. This is business. This is how I make my living.

DO FARMERS MAKE APPOINTMENTS?

They would, if you gave them the chance. Actually, the more the agricultural market changes, the more willing farmers are to make appointments. In fact, the new farmer is insisting that the salesperson make appointments. Why not? This is the way business is conducted everywhere else. In today's world, appointments are no longer options. This is not to say drop-ins are no longer welcome. In the right relationships, they can be very effective. However, sales calls can and should be done by appointment.

There are still some farmers who prefer not to work by appointment. Never assume this. In fact, you should assume the opposite until you find out to the contrary. How do you find out? Just ask!

"Mr. Farmer, should I just drop in at any time or do you prefer me to set a definite time with you before coming out?"

You may also want to set appointments for the farmer to come to the store. Again, you could ask if it would be better for you to come to them or for them to come to the dealership. Always add this:

"If you do, I'll make sure we're not disturbed when you come in."

WHY MAKE APPOINTMENTS?

Why make appointments at all? Besides the fact that it's what the customer is coming to prefer, it saves you time, gives you a definite purpose when calling and prepares the farmer for your call. Here are some other compelling reasons.

Making an appointment tells the farmer:

1. You mean business.
2. You are a professional.
3. What you have to say is important.
4. You value his time.
5. Your time is valuable, too.
6. You have a very specific reason for meeting and that is to sell him something. There is no deception.

WOULD YOU PREFER TO HAVE APPOINTMENTS?

The answer is obvious when you consider the value of appointments. Yet many agricultural dealer salespeople still shy away from appointments as if they were undesirable. This is changing as more and more salespeople come to realize just how little time they have to waste.

P.A.S.S. C.A.L.F.

DO FARMERS KEEP APPOINTMENTS, AND IF NOT, WHY NOT?

Whenever somebody does not keep an appointment with you, it's perfectly natural to be disappointed. It's also normal for optimistic salespeople to find excuses for the prospect. They "understand" how busy the farmer is or that he could easily forget. There are legitimate reasons why people miss appointments. A death in the family or a tornado qualify as legitimate reasons.

Most times the blame for a missed appointment can be laid squarely on the shoulders of the salesperson. Why? Look at the way you make appointments and the things you say in setting them up.

Do you use the alternative close? "Would 9:14 a.m. work for you or would 1:17 p.m. be better?" If you do, consider this: It's amateurish. So many salespeople still use it, even though customers long ago figured it out as a sales trick.

Do you tell your prospect you will be in the area and could you drop in while you are there? If you do, you have minimized the importance of the call.

Do you avoid telling the prospect why you want to call? If you do, you are setting up a false expectation that can lead to customer frustration.

Are you deliberately vague about why you want to meet? If you are, you have not given the prospect a reason for meeting you.

Besides being amateurish and ineffective, these behaviors detract from the importance of the meeting. If it's not important, don't waste each other's time. Life is too short. Time is too valuable.

MAKE THE APPOINTMENT IMPORTANT

If you're going to make appointments that people keep, make the appointment important. In most cases, salespeople have to travel several miles to a farmer. This takes valuable time. If he is not going to be there, it's wasted time. Even when he says he is coming in to the dealership for the appointment, you have to set aside time that is no longer available for other things if he doesn't show. This, too, wastes time.

How can you make the appointment important? What you say to the prospect and how you say it can determine in the prospect's mind how important it really is. Consider the following scripts.

> *"If you come into the dealership, I'll make sure we're not disturbed. I'll put a hold on all calls from 2 p.m. until 3 p.m. so I can devote my full time to you. If I'm out in the country, I will make sure I get back by 2 p.m., even if I have to miss out on other calls. If you're going to be late for any reason, will you please call me so I can adjust my schedule? Even if I'm not in, the person in the office will get your message to me wherever I am."*

> *"I'll come by your farm and bring some brochures and other information on that tractor. I'll be able to answer all the questions you may have to make an informed decision about it. This will give us an opportunity to properly discuss it without any interruptions. I can be there Wednesday at 9 a.m. Will you have some coffee ready? I'll bring the donuts. We can sit down and see if this tractor can do the job for you*

P.A.S.S. C.A.L.F.

without any pressures. Is that okay with you? Now, I'm coming out especially to see you. It's about an hour's drive from the dealership, and I have no reason to come out there except to see you. If for any reason you can't be there, or you will be late, can you please call me before I leave the dealership?"

Can you see how what you say and the way you say it can affect the way the customer will feel about the importance of the meeting?

Many salespeople are afraid to be that assertive. They feel the customer may not be comfortable with this approach. They fail, however, to look at the opposite approach and how absurd it is when held up to objective criticism. They still prefer deception and intrigue to get appointments. Whenever they are successful at it, they point it out triumphantly as if to say, "I told you so!" They forget that this approach also results in many appointments either not being kept or being totally ineffectual. Business today is a whole lot more straightforward than that.

SOME ADVICE ON MAKING APPOINTMENTS

1. Make the appointment important.
2. Build the rest of your day around your appointments.
3. Keep some elastic time. Don't fill your day with appointments to the point that you are unable to take care of other, equally important tasks.
4. Confirm appointments made more than a few days in advance.
5. Call while on your way to remind the prospect about the appointment.

6. Use the phrase "let's sit down" to avoid standing in a field in the rain.

WHAT IS THE BEHAVIOR?

Make at least one selling appointment per day. This appointment must be with someone you have a chance of selling something to that day. Do not go to bed unless you have one of these appointments set up for the next day. Preferably, make these selling appointments a week ahead.

WHY IS MAKING SELLING APPOINTMENTS A SUCCESSFUL SELLING BEHAVIOR?

Making selling appointments is a successful behavior that will keep making money for you. Most salespeople have several appointments in a day, and they will tell you all appointments are selling appointments. They like these word games. They are not always sure which of these appointments will result in actual sales. Practice this behavior and you will guarantee sales well into the future. Will you sell every one of them? No. But you will have a better chance of making sales consistently than the other salespeople in your dealership. This behavior requires you to do something consciously. Once you practice it for a few weeks, it will become a habit that eventually ends up being the way you do business. Ask any salesperson if he or she would like to have one legitimate chance of selling someone each day. Ask yourself.

P.A.S.S. C.A.L.F.

IF YOU ARE NOT AN AGRICULTURAL SALESPERSON

Once again, this chapter is not limited to agricultural salespeople. It has universal applications. Having one selling appointment for each and every working day can dramatically affect the amount of money you make. You may be in a business that does not provide such opportunities. For example, you may sell very large items that take years to close. This behavior still applies to you. In complex sales, there are many small sales along the way. If you practice the behavior in this chapter, you could shorten your sales cycle. In complex sales, this behavior becomes even more important because it forces you to strategize more effectively. It helps break the big sale into smaller components that also require selling.

If you are in other "normal" sales jobs, this behavior could make or break your career. At the very least, it can make the difference between mediocre earnings and superstar earnings. Which would you prefer?

BEHAVIOR #3

SOLVE PROBLEMS

WHY CUSTOMERS BUY

Why does a farmer buy equipment? The answers given by agricultural salespeople in my workshops point out an interesting dichotomy. They generally give emotional reasons such as "keeping up with the Joneses," or practical reasons such as using it for tax purposes. They rarely say, "Because it is made of high quality steel," "It has rubber tires," or "It has a short turning radius." They avoid talking about the features of the equipment in a classroom setting. Yet in the field, I have observed them doing just that. They bombard the farmer with the marvelous attributes of their tractor. They talk proudly of the quality of their equipment—without translating these benefits into practical or emotional reasons for the farmer. So much of what they tell the farmer is information-bound. They forget the real reasons the farmer will buy.

This behavior asks you to answer a very simple question: What problems will your equipment solve for the farmer? It also

P.A.S.S. C.A.L.F.

asks you to first define the problem and then find a solution. It asks you to look at things through the eyes of your customer.

Few salespeople are unaware of the value of solving problems. Classroom responses suggest they are very much aware of the importance of this behavior. They simply ignore it in the field. Why? Perhaps they are more comfortable explaining features and benefits. Perhaps it's what they have been taught. FAB—features, advantages, benefits selling—is outdated. However, many agricultural salespeople still cling to it. They are unable to incorporate problem solving into their presentations.

I was fascinated by a survey conducted by a marketing consulting company. (I apologize to the company for not giving it proper credit. I simply cannot remember where I read this or who the company was.) They had asked customers for the attributes they would most like to see in their salespeople. They then asked which of their salespeople had those attributes.

This is where most research stops. It is assumed that the customer has given us pearls of wisdom and we should fashion our salespeople accordingly. But this company went further and looked at more objective information. They looked at how many purchases these customers had made that would not have been made had the salesperson not possessed these wonderful attributes. They found that, of all the things customers said they like in their salespeople, only one made an actual difference in the number of purchases they made. This one attribute, simply stated, is this: "He knows my business."

This may seem to be one of those self-evident truths. Yet the number of salespeople who fall woefully short of the mark when it comes to this simple standard appalls me. Learning your customers' business is one of the best uses of your time. It pays big dividends. And it's essential in order to make problem-solving work. How can you solve a problem if you cannot identify the problem in the first place? How can you identify a problem if you don't know the business from your customer's point of view?

When salespeople start to think in terms of solving problems, when they start to see things through the eyes of their customers, they move from simply selling equipment to becoming a resource and a consultant to their customers. This is a very desirable position to be in. It forces the customer to rely on you and your advice, and it separates you from your competition.

If you ask salespeople what the biggest thrill of doing their job is, you are likely to get many different answers. For example, my biggest thrill in a sale comes at the moment when I know the prospect has bought but has not yet realized it himself. But there is one aspect of selling that is universally accepted as being very high on the "most gratifying" list. This is when the customer looks to you for advice and relies on the advice you give. Most salespeople strive to get to this position because it means they have elevated themselves to the level of the true consultant. Solving problems is one way to catapult you to that position.

Some "do's" and "don'ts":

1. Do stop thinking of how wonderful your equipment is or how many features it has.
2. Do stop thinking about equipment and what equipment can do.
3. Do start thinking about what your equipment can do for your customer.
4. Do look for a problem first and then find a way to solve it.
5. Do ask yourself if this is important and why.
6. Do ask yourself what need it will fill.
7. Don't tell the farmer what the equipment can do for him – ask him.

P.A.S.S. C.A.L.F.

LOOK FOR PROBLEMS

What problems do your farmers have? That's where this behavior starts. Once you have identified the problems, ask yourself if your equipment can resolve all or at least some of them. Are they important or urgent problems? How important or urgent are they, and why? What might happen if your farmer did not buy your equipment? What might happen if he did?

The answers to these questions force you to look at the farmer's business through his eyes. Think of yourself as a problem-solver, not just another salesperson trying to make quota. The results may startle you.

$4 MILLION OF NOTHING

A friend who sells consulting services called me excitedly. He had a meeting scheduled with a major prospect the next day. As far as he was concerned, the sale was in the bag. He was meeting with some bigwigs at JC Penney. When I asked why he was so confident, he told me he was going to save them $4 million! "It just makes so much sense," he told me.

"Besides the $4 million savings, why else should they buy from you?" I asked.

"Are you kidding?! That's four million reasons!"

"You'd better find another angle," I suggested. "Saving money is not that powerful a motivator."

He called me the next day.

"What happened?" I asked. "Do I congratulate you?"

"No." He sounded despondent. "Do you know what happened when I told them I could save them $4 million? One person actually yawned!"

He proved once more that one should find a problem, determine how important the problem is, and then find a solution. For a company the size of JC Penney, $4 million is not even money. It's all relative. There had to be other, more compelling reasons for them to change what they were doing. My friend learned that the hard way.

Go one step further. Learn to anticipate problems that could arise in the future. Help the farmer plan to avoid these problems or at least minimize them. You can only do this when you know the farmer's business.

WHAT IS THE BEHAVIOR?

Solve problems, don't just sell iron. See things through the eyes of your customer. Learn your customers' business well enough so you can become a consultant and advisor. Look for problems they could have and find solutions for them. Sell solutions, not iron.

WHY IS SOLVING PROBLEMS A SUCCESSFUL SELLING BEHAVIOR?

This behavior requires action on your part. You cannot passively become a problem-solver. You must do something. There are several action steps to this behavior.

1. Identify who your customers are.
2. Actively inform yourself about their business.
3. Actively look at their business as if it were your own.
4. Pre-empt problems and bring them to the attention of the farmer.
5. Actively look into the future with the farmer to anticipate future problems and find solutions before they are needed.

These are not things that will just happen. You must make them happen. If you are content to just sell iron, then becoming a problem-solver may not be a priority for you. However, if you want to guarantee that your customers will always need you and will remain loyal to you, then you ignore this behavior at your peril!

The survival of your dealership—and your very job—depends on your ability to make this behavior real for you. It is becoming increasingly more important for all agricultural salespeople to understand and practice this behavior.

Like the other behaviors, if you apply it and perfect it, you will make it a habit. Solving problems can become habitual for salespeople. When it does, it becomes the way you do business. It also increases your value to your customers.

IF YOU ARE NOT AN AGRICULTURAL SALESPERSON

Take out the agricultural terms and you will see this is a behavior that applies to all industries and all selling. Why shouldn't you think this way? Isn't this what your customers expect from you? Whether you sell dishwashers or pagers, people buy from you because of the value they perceive you bring them. You are not selling anything intrinsically different from your competitors. This is one way to differentiate yourself and keep customers coming back to you.

How much is enough knowledge of your customers' business? Only you can tell. Look at things through the eyes of your customer and ask yourself how much knowledge you would like your salesperson to have if you were in his shoes. If you don't already possess this knowledge, you should be doing something about it *right now*.

In some industries where competition is fiercer, this behavior becomes even more important. It doesn't matter how much or how little your product or service costs. Your customers often expect you to push only what you sell. Surprise them by showing how much you know and how this knowledge can help them make better purchase decisions.

BEHAVIOR #4

SIX PROSPECTS

WHAT IS ASTRONAUT EQUIPMENT?

Equipment on your lot that takes up space for years and then ends up in the Black Hole (on the auction block)...otherwise known in the industry as "lot rot"...equipment you have that you wish you had not traded in...stuff you don't even care to look at.

However you slice it, astronaut equipment is expensive. This is a subject most dealers would prefer not to talk about, even while they wish they could find a solution. They blame everyone else—even neighboring dealers—for these undesirable additions to their lots.

HOW EXPENSIVE IS ASTRONAUT EQUIPMENT?

Most dealers tell me that having this equipment on hand is not like money in the bank. It does not accumulate interest. In

45

P.A.S.S. C.A.L.F.

fact, they pay interest on it. Most accountants can tell you, to the penny, how much. They could even give you a daily tally as the costs mount. These are only part of the costs, however.

When the farmer traded that piece in, the salesperson taught him a very valuable lesson: *Next time I have a piece of junk, I know where to find a home for it.* The farmer has learned he can trade in junk to this dealer, even if other dealers will refuse to take it.

You'd think he would at least be grateful. You'd think that later, when he has a good trade, he would come looking for the salesperson to reward him for his generosity. Unfortunately, the opposite tends to happen. Thinking the salesperson will use the next opportunity to recover some of his losses, he takes his good trade to a neighboring dealer. This represents a lost sale.

Other costs are even more subtle. The farmer and his wife drive to town and see the astronaut equipment on your lot. They think nothing of it. They drive home and see it again. Again they think nothing of it. However, after this happens several times they start to associate bad equipment with your dealership. They don't do it deliberately or maliciously. It just happens. The association chips away at the positive image you are trying to build. That costs.

Most dealers readily admit this equipment is expensive. They also admit they would rather not have it. Many will go so far as to confess they have made some bad trades. So why do salespeople continue to make these bad trades? Is it because they are stupid? I don't think so. I know Yielder™ Call Reluctance® plays a big part (see sidebar, page 92). Perhaps it is also because they have not found a better way. The Six Prospects behavior may be part of the solution.

WHY DO DEALERS ACCEPT ASTRONAUT EQUIPMENT?

When asked why, most dealers either shrug their shoulders or attempt to justify their actions. They spend an inordinate amount of time finding reasons why they did the trade. Most of these reasons sound like lame-duck excuses. Some blame their salespeople, even though they gave the salespeople the authority to drag the junk in. Whatever the reasons they give, they all boil down to one of two things: They were afraid they would lose the sale if they did not take the trade, or they could not say no to the customer.

The first is a reflection of the intense competition among dealers to land a deal, to sell more iron. It implies that the only way to make money is to make unprofitable deals. If I don't do it, they think, then my competition will. This may even be true in some cases.

The second is a lack of assertive selling. Dealers want the business and want to show the customer they will do anything to get it. In this way, they hope to gain the respect and affection of the customer. This blithely ignores a simple truth. In most cases, the customer knows he has astronaut equipment and is looking for a sucker to take it and pay him for it. When you do, he may like you but he certainly will not respect you. Think about it. Would you respect someone who is a bad businessperson? Is this the person you would rely on for help and advice when you need to solve problems?

Take enough pieces like this and your dealership eventually will become the junkyard of the community.

P.A.S.S. C.A.L.F.

HOW DEALERS COPE WITH ASTRONAUT EQUIPMENT

Some dealers become very creative in the way they cope with astronaut equipment. One dealer I know refuses to keep any used equipment at all. "If I am going to lose money," he says, "it will be on new, not used, equipment." His solution is to turn all used equipment over to jockeys immediately. This limits his losses and helps him to avoid looking at the stuff. But his lot looks bare.

Another dealer waits until he can stand it no more. Then he goes around the lot, painting a red mark on any piece he considers a loss. He calls the junkyard and has them come and haul those pieces off. This is creative. Expensive, but creative.

But the solution of choice is the infamous auction sale. Most dealers use this as a way to clear their lots of astronaut equipment at a loss. None of the dealers I know tell me they make any profit on these auction sales.

Lucky is the dealer who does not have astronaut equipment.

WHAT IS THE SOLUTION?

The solution could be a behavior I call Six Prospects.

Before you trade in *any* piece of equipment, ask yourself three questions:

1. Do I know six prospects who are likely to buy this piece?
2. How much are they likely to pay me for it?
3. Why should they buy it? What can it do for them?

If you do not get the right answers, you know two things:

1. This is not a popular piece of equipment.
2. You are looking at astronaut equipment.

Does that mean you will not trade? Not necessarily. You may still want to trade based on other factors. However, you will trade with your eyes open. Perhaps you want to buy the business. Perhaps you won't mind losing some money to convert another color customer. Perhaps the deal is big enough for you to recover your losses in some other area. Perhaps you're just feeling charitable. There could be any number of legitimate business reasons for doing the trade anyway. At least you will know ahead of time. At least you won't display this piece prominently on your lot. (If you can't think of six people who might buy this piece, how likely is it that someone will drive by and say, "Wow! That's exactly what I'm looking for!"?) You can at least avoid the association aspect I mentioned earlier.

If you do get the right answers, you should call your six prospects. Maybe that piece will be sold before it even hits your lot!

Maybe, but many salespeople tell me this is not always possible. They remind me that "when in the heat of the sale…" My response to this is simple. I have been in the heat of the sale more times than I can remember. The lesson I have learned is this: He who loses his head in the heat of the sale loses money. The heat of the sale is no excuse for poor judgment. Salespeople who use this excuse should seriously look at their careers. Their sales managers should consider whether sales is the right career for them.

THAT *#*@* SALES MANAGER JUST COST ME ANOTHER DEAL!

It was a cold autumn morning. The agricultural salesperson (call him John) and I stood in a drafty barn. He was trying to sell a new tractor to the farmer. I was along for the ride.

The farmer had agreed it was exactly what he needed. But he wanted to trade in a used piece and wanted an exorbitant amount for it. John tried hard to get him to reduce his price, but the farmer held on. I could see the farmer was relishing this game. There was a twinkle in his eye even as he told John how disappointed he was they could not reach an agreement.

John really tried. He tried even harder to convince his sales manager back at the dealership to let him make the trade. Four times he went to his pickup to call his sales manager. Four times he came back, red-faced and obviously frustrated. His sales manager would not budge. He knew John was looking at astronaut equipment. Even I, with my limited knowledge of farm equipment, knew that just from looking at it. I was surprised John did not see it. He was livid. Finally, in desperation, he told the farmer he could not do the deal today. He would go back to the dealership to talk directly with his sales manager and get back with him the next day. The farmer looked genuinely disappointed but told him to take his time.

As we walked back to his pickup, I could see John was angry. "This is not the first time that miserable s.o.b. has cost me a deal," he muttered angrily. He then went on to use some expletives that described his sales manager and his family. As we were about to get into the pickup, I asked timidly, "What would happen if you went back and told the farmer you didn't want his equipment and asked him to buy yours anyway?"

He looked at me in disbelief. "You don't understand," he said in a voice too ominous to be casual conversation. "He wants to trade that piece in and he won't buy unless I take it."

We were on a strange farm, miles from the dealership. John was my ride back, so I was not about to upset him even further. We got in and rode back to the dealership in silence. I was deep

(continued)

> **THAT *#*@* SALES MANAGER** (continued)
>
> in thought. I had just learned something new about agricultural salespeople. They could read customers' minds! Not only that, they could even make up their minds for them!
>
> The next morning, while I was waiting to go out with another salesperson from the same dealership, we heard that the farmer had bought the same piece of equipment John had tried to sell him. He had bought it the night before from a neighboring dealer – and he had not traded in anything.
>
> I was tempted to go to John to say, "I told you so!" But he was twice my size and I didn't think it would be a wise move. Some other salesperson had seen the opportunity and stood his ground and made the sale. Pity, it could have been John.

OTHER POSSIBLE SOLUTIONS

Frequently the farmer knows his equipment is not worth much and is simply trying to offload it. You can't blame him for that. It's good business from his point of view. That doesn't mean you should accept it. What can you do? Ask him whom he knows who would likely buy this piece and how much that person is likely to pay for it. Whatever he answers—and even if he does not answer at all—at least you will have succeeded in getting him to become more realistic about the price in his mind.

You may even refuse to accept the trade altogether and ask him to buy your piece anyway. Even though this sounds risky, it may be what is required to get the sale completed on the spot. (See sidebar for an example.) Who knows? He may be as anxious as you are to get the deal closed.

P.A.S.S. C.A.L.F.

> ### SALESPEOPLE ARE BUSINESS PEOPLE, TOO
>
> If you're in sales, you are in business for yourself. Most salespeople agree with this statement. Yet they do things that would bankrupt most businesses. They forget the role of a business person. When you are in business, you are both a buyer and a seller. You must "buy" some things your customer is selling—objections, price negotiations, and of course the idea that "this will benefit you more than me." Your reaction to these "transactions" determines how long you stay in business. A good business transaction is one that benefits all parties. Sometimes salespeople forget that.

WHAT IS THE BEHAVIOR?

Whenever you are faced with *any* trade, ask yourself these three questions:

1. Do I know six prospects who are likely to buy this piece?
2. How much are they likely to pay me for it?
3. Why should they buy it? What can it do for them?

Then call them!

WHY IS SIX PROSPECTS A SUCCESSFUL SELLING BEHAVIOR?

This behavior requires action on your part. You cannot passively make it happen. It requires several other actions, as well. You should already have a profile base and have learned enough about your customers to be able to select six prospects.

Six Prospects requires you to think and keep your head in the heat of the sale. You must actively develop a different attitude toward business. Instead of grabbing what's in front of you for the sake of making a sale, you must become more aware of the implications of the sale. Will you sell less iron? Probably not. You may even sell more, and at a profit. There is nothing wrong with making a profit. Customers expect you to, no matter what they tell you.

Sometimes refusing to lose money on a deal may just get you what you really want: respect from the customer. If you take this behavior and make it a habit, you will make fewer trade-in errors and sell more new and used equipment. It should be the way you do business.

IF YOU ARE NOT AN AGRICULTURAL SALESPERSON

If you do not take trades in your business, this behavior may be more difficult to apply outside the agricultural world. However, many customers have to get rid of something in order to purchase something else. By practicing this behavior, you may make it easier for them to do business with you. Take out the agricultural terms and see whether it works in your business.

Like the previous behavior, Six Prospects requires you to know your customer's business. If you can help or advise your customer about selling his or her trade, you become a valuable resource – even if you don't take trades yourself.

Try Six Prospects, and you may find yourself selling into the future as most top salespeople do.

SELL INTO THE FUTURE

When I was president of Sales and Marketing Executives of Dallas, I organized a panel for one of our annual seminars. This panel consisted of sales superstars who were recognized as the best by their peers. Salespeople from many industries filled the auditorium at Southern Methodist University to ask questions of these top sellers. I listened carefully because these were classy people. I also made copious notes. Reading my notes later, I noticed that each panel member had agreed on one sales principle. Each applied this principle in his or her own unique way, and I'm sure they did not even consciously know they were doing it.

What each one said, in different ways, was this: "I don't just take the sale in front of me. I always look for additional sales for the future."

In other words, they practice a principle I teach called "unfolding"—selling for the future. The Six Prospects behavior starts salespeople toward selling for the future.

BEHAVIOR #5
COLD CALLS

WHAT IS A COLD CALL?
WHAT IS A COMPETITIVE OWNER CALL?

Probably the most misunderstood terms in dealerships are "cold calls" and "competitive owner calls." I have heard more variations on the meanings of these terms than any others. Therefore, it's appropriate to give Sales Academy's definitions as a prelude to this chapter.

Cold Call – This is any call, whether in person or over the telephone, on someone who has not done business with your dealership at all or within the last year.

Competitive Owner Call – This is any call, whether in person or over the telephone, on someone who has not done business with your dealership at all or within the last year.

P.A.S.S. C.A.L.F.

Did you notice they are one and the same thing? Those interminable arguments over what constitutes a competitive owner become meaningless when you consider this fact: If a customer is not buying from you, he is buying from a competitor. That makes him a competitive owner. Whether this competitor is yours or another color makes no difference. He is not buying from you. It doesn't mean he is not buying at all. If he is a potential customer, then he is buying – just not from you. This simplifies things somewhat. It makes every cold call also a competitive owner call.

Because I cannot accept that a dealership in a community has not called on every single farmer in that community at least one time, I will expand the definition of a cold call/competitive owner call to include at least one of the following:

1. Someone you have not personally contacted before.

2. Someone who has never done business with the dealership before.

3. Someone who has not bought anything from the dealership in the past year.

4. Anyone else who is buying farm equipment, but not from you.

After you have exhausted all the above, you may want to redefine cold calls. You could alter the definition to mean someone you have not contacted in the last six months, or whatever other criterion you choose.

> ### COLD CALLING IS LIKE BUYING
> ### A LOTTERY TICKET!
>
> My son, Ade, is a physicist and mathematician. It drives him crazy when I buy lottery tickets. He once asked me if I knew what the odds were of my actually winning the lottery. "Yes," I told him, "it's 50-50. I either win or I don't. I also know what the odds are if I don't buy a ticket." He threw his hands up in despair. Even though his other major was Philosophy, he was unable to counter this argument. The same holds true for salespeople making cold calls. You'll either make a sale or you won't. Fifty-fifty. Not bad odds. Much better than the odds if you don't make the call.

WHY MAKE COLD CALLS?

"Cold calls" don't even sound nice. Most salespeople, no matter what they sell, don't like making cold calls. After all, the risk of rejection is much higher than with a warm, friendly face. There are salespeople who tell me they love making cold calls. I think they either have elephant skins or they stretch the truth. There is nothing especially inviting about cold calling. It's a chore at best. So why bother? Especially when you already have enough business?

Perhaps one answer is that the business you already have may be fragile. Your competitors are either contemplating taking it away from you or have already made some effort to do so. Since no business can be relied on to continue forever, you may be well advised to keep adding to your list of customers.

Another reason could be that you would be doing something your competitors routinely avoid. If success in sales really does result from doing the things others do not like doing, then cold calling is definitely for you.

HOW IS IT POSSIBLE?

It was after a workshop I had just conducted for some agricultural dealer salespeople and sales managers. Bill (not his real name), a wily, hardened dealer principal, cornered me.

"You talk about profiles as if you know so much about it," he accused me. "Yet you don't realize what it takes to get a profile."

"I'll be the first to admit I don't know everything there is to know about profiles," I agreed. "But what is your comment? Are you saying it is impossible to collect profiles?"

"No, it's not impossible. It's just not as easy as you make it out to be."

"So what you're saying is that it's not impossible; there is a process."

"Darn right!" He knew he had me.

"So what's the process?"

"Well, you first have got to get to know the customer before he will even agree to let you profile him. This takes time."

"Okay. So getting the profile is possible as long as you follow the process, and the process requires you to get to know the customer well enough. I'll buy that. How many times do you need to call on him before you know him well enough for that?"

"Depends on the customer. Three, maybe five times. Sometimes as many as ten calls."

"So what you're saying is that getting the profile is possible as long as you call on the customer at least three times, maybe as many as ten times. Then the customer will allow you to profile him."

"Now you're starting to understand." He smiled approvingly. He had just taught this know-it-all a lesson.

"I'll buy that, too," I responded. "I take it you have not collected many profiles?"

"No. I told you how difficult it is."

"You obviously know your business very well. Tell me – how long have you been in business?"

He flashed me a smug grin. "Twenty years!"

(continued)

> **HOW IS IT POSSIBLE?** (continued)
>
> "Are you telling me that, in twenty years of being in business, you have not called on your customers at least ten times?"
>
> The grin disappeared. He stared at me for a while and then abruptly turned and walked away. I was not to be denied. Grabbing him by his shirtsleeve, I persisted.
>
> "How can it be possible? You live in the country. It's not as if you live in Dallas, where there are four million people. You have a finite number of customers. How can you not have called on them ten times in twenty years?"
>
> He mumbled something about me not understanding and went off in a huff. This was not a good way to win friends and influence people.
>
> This question has bothered me in my encounters with dealers. How can it be possible for them to even talk about cold calls? Surely, in a rural community where your livelihood depends on farmers, you would have called on every farmer in your community several times. Bob Honzik, a former Division Manager and now happily retired, called it "running the same old trap lines," and he was right. It is so much easier to keep calling on the same people and to ignore the rest of your market.

There are other compelling reasons for making cold calls.

1. You may just sell something!
2. You keep your selling skills sharp.
3. You exercise your skills by getting out of your comfort zone.
4. You become known in your community.
5. You guarantee a constant source of new business.
6. You get to your competitor's customers.
7. Cold calls will drive your competition crazy—but don't worry. They won't retaliate. They're as

P.A.S.S. C.A.L.F.

terrified of making cold calls as you are. But you can do it.

The successful behavior here is to make at least one cold call each and every day, no matter what. There are busy times of the year when this becomes a real chore. It doesn't matter. *Do it.* This behavior will keep you in business long after your competitors have up and left. In less busy times, make more than one each day.

A STUDY IN INDIFFERENCE

Back in 1992, one of the big agricultural equipment manufacturing companies did a study. They had teams call on competitive owners, competitive dealers, their owners and their dealers. The results alarmed them. They found that competitive dealers were calling on their customers more often than their dealers were, and that their dealers were not calling on competitive owners at all.

HOW DO YOU MAKE COLD CALLS?

Even after salespeople are convinced that cold calls can help them, they still hesitate to actually make them. Some will say they simply don't know how. Since most basic sales training classes teach you how to make cold calls, this is hardly an excuse. Most times when salespeople avoid cold calls it is the result of unresolved cases of Sales Call Reluctance®, not lack of knowledge. (See sidebar, page 92)

Perhaps the following scripts can help.

"I'm calling because we have never met and most people around here already know me. Can I stop in and meet you sometime? I'd just like to shake your hand and leave you with my business card. Tell you what—I'll be in your neighborhood next Friday around ten. If you have some coffee ready, I'll bring some donuts and we can chat for about fifteen minutes. Is that okay with you? If something comes up, can you call me before I buy the donuts?"

"Hi. I can't believe there is still someone in this community I haven't shaken hands with, so I stopped by to do just that. My name is..."

ONE-TRIAL OR ZERO-TRIAL LEARNING?

There were eight salespeople at the dinner table with me in Minneapolis. All had just completed the Fear-Free Prospecting and Self-Promotion Workshop®. Conversation was animated. They were enthusiastic about the workshop they had been through and ready to put into practice some of the things they had learned. I was engaged in conversation with Billy, who sat to my right. The topic of competitive owners came up.

"What's your success rate with competitive owners?" I asked him.

"Oh, about one in thirty, I guess," he replied.

The others heard my question and volunteered their opinions. They quickly agreed that calling on competitive owners was not very productive. In the end, the agreed success ratio was one in thirty.

"Wow!" I was thrilled. "Does that mean you actually called on thirty competitive owners?"

"No. But if I did..."

Maybe this was real experience talking. Maybe it was what author George Dudley calls the Principle of One-Trial Learning. Or, even worse, the Principle of Zero-Trial Learning.

P.A.S.S. C.A.L.F.

WHAT IS THE BEHAVIOR?

Make at least one cold call each and every day you are in business, no matter what. Make more when time permits. Don't go to bed without making that one cold call.

WHY IS COLD CALLING A SUCCESSFUL SELLING BEHAVIOR?

Cold calling is a behavior that requires action on your part. You cannot passively make it happen. It requires several other actions, as well. You must already have learned the names and addresses of potential customers you have either not gotten to see or have avoided. You must have developed a planned approach to contact all of them within a certain period of time. You will already have placed their names in a special file that corresponds to each area in which you call so you can select names easily. You must also have consciously decided that you will make that one call each day, no matter what.

This is a difficult behavior because it is not as pleasant or easy as some of the others. However, it will keep you busy when others slow down, and it will provide a constant source of new business while others complain. In the agricultural world you will be on your own. If the studies done are correct, you will have very little competition from other salespeople, especially if you call on other colors. Paradoxically, of all the markets available to you, this is the most lucrative and the one least filled with competition.

Make these calls every day until it becomes a habit. Then it will not be simply the way you do business but the way you grow your business.

IF YOU ARE NOT AN AGRICULTURAL SALESPERSON

Once again, take out the agricultural terms and this behavior applies to you. No matter what you sell, no matter how busy you are, making that one cold call each and every day will separate you from your competitors. If your business is making cold calls, then make one more than your peers do.

Salespeople in all industries fear cold calls, even though intellectually they know there is nothing to fear. I admit this is one of the least favored of all sales activities. That is the most compelling reason to make them. If you feel queasy about making cold calls, chances are your competitors feel that way, too. Success in sales, as in most other endeavors, comes from doing the things others avoid. Cold calling is one of those things.

Will you sell every time you make a cold call? Of course not. But when you do, it will usually be the most profitable and gratifying sale that day.

BEHAVIOR #6
ASK WHY

FOUR LEVELS OF SELLING

Of all the behaviors, this is the easiest to understand in concept and the most difficult to put into practice. It is also the one behavior that can, by itself, raise your level of selling no matter where you are. It is the one behavior the sales superstars of the world do automatically.

I believe there are four levels of selling.

PRODUCT SELLING
Some salespeople sell in terms of product only. This limits their ability to sell because their presentations tend to be boring and not very effective.

FAB
Some sell in terms of features, advantages and benefits. This increases their ability to sell because they have progressed beyond the product stage. Their presentations are likely to

P.A.S.S. C.A.L.F.

be livelier. However, they may tend to revert to the product stage because all the features, advantages and benefits are, after all, tied to the product.

ME

Some salespeople still believe the customer buys because of them only. This is the "Me Syndrome." They usually experience some success because they tend to develop good relationships with customers. However, they are limited in their ability to sell because they believe the only product they are selling is themselves. They tend to become self-centered and easily distracted when a customer does not show enough interest in them and their abilities.

BEYOND FAB

A higher form of selling is what we call "Beyond FAB." This goes beyond the old-style features, advantages and benefits method. It focuses on the customer and tries to determine *why* the customer should buy from the salesperson. It focuses on solutions to real problems. Very few salespeople reach this level without proper training. It assumes nothing as far as the customer is concerned and uses strategic questions to determine how best to sell to the customer.

The Ask Why behavior takes salespeople to the "Beyond FAB" level of selling. It uses questions to uncover needs and to sell. It asks "why" twice—when the customer is about to buy and after the customer has bought. It constantly uses the right questions to make a sale that sticks. Most salespeople learn to ask questions in basic selling classes. However, most sales training stops at this point. Ask Why picks up where other selling programs leave off and uses questions in a more ingenious way.

Let's first look at questions in a sales setting.

QUESTIONS IN SELLING

There are many types of questions in a selling situation. Each has its place.
1. Information-gathering questions net you information you do not know, such as a mailing address.
2. Validating questions are used to validate or disprove data.
3. Attitude questions uncover values and feelings.
4. Closing questions are used to close or trial-close a deal.
5. Finally, there are "Ask Why" questions designed to lead, direct, inform and sell.

Since this behavior requires us to use only "Ask Why" questions, I will deal only with these.

WHAT DO "ASK WHY" QUESTIONS DO?

Selling questions are unlike questions in a social situation. In a social setting, I may ask you how your family is. I have no way of knowing what you will answer. You could give me a very short "fine" or you could spend the next two hours crying on my shoulder. Your answer could surprise me.

In sales there are no such surprises. You will never ask a question you don't already know the answer to.

That may sound strange, so let me repeat it in case you think it's a typo. *In sales you will never ask a question you don't already know the answer to.*

Whenever I tell this to salespeople, they usually look at me as if I had taken leave of my senses. All the credibility I have

P.A.S.S. C.A.L.F.

built up with them goes down the toilet with one statement. And then I explain and they understand and I regain my lost credibility. So let me explain.

Unlike other types of questions in a sales setting (such as information or attitude questions), the questions used here are not designed to elicit information you don't know. The purposes of "Ask Why" questions are to lead, direct, educate or inform the prospect, explain a point, ensure understanding, and sell. Because you do not need to gather information, the questions you use will instead get prospects to answer their own questions.

How can you be so clever? The question you ask is really a statement in the form of a question. You may not know the exact words you will get in response. However, if you get an answer that does not conform to what you know it should be, then one of two things is happening: Your prospect did not understand the question, or you're talking to the wrong person.

As an example, if you were to ask a farmer why he wanted to buy a combine and he told you he wanted to use it to go fishing, you would know something was wrong.

"Ask Why" questions uncover need in the strongest way possible. They help prospects discover their own needs and answer their own objections. "Ask Why" really asks why customers want what they say they want and then asks why they bought it. "Ask Why" questions always require answers.

EXAMPLES

Perhaps a few examples are in order here. Notice three things as you read them.

↓ There is always a "why" to the question.
↓ The question requires an answer.
↓ You already know what the answer should be.

"What do you want that for?"
"Why would you want to do that?"
"If you had this tractor, what would you do with it?"
"Why?"
"What would happen if you did not buy this tractor?"
"Would that hurt?"
"What would happen if you had a bigger or newer tractor?"
"If I could have one on your farm in the next two days, would that help you?"

DON'T STOP WITH WHY

Don't stop with the "why" questions. Always follow them with "importance" questions:

"Is that important to you?"
"Why?"
"How important is it?"

A simple sequence of questions goes like this:

"Why do you want to buy that?"
"Is that important to you?"
"How important is it?"
"Why?"

P.A.S.S. C.A.L.F.

You should develop your own set of "why" questions. Develop them for every piece of equipment you sell, every problem you encounter and solution you offer.

WHY ASK WHY BEFORE THE SALE?

If you ask "why" questions before the sale is made, you increase your chances of making the sale. Instead of telling prospects why they should buy your tractors (don't you just love it when salespeople tell you why you should buy something?), you ask them to tell you. You become a sounding board. By using the right questions, you can lead them to sell to themselves. There is no salesperson as powerful as the person who is buying. So why not use the most powerful selling tool you have? Let your prospects sell you and themselves on why they should buy from you.

Most salespeople cannot get this far. After all, they know all the answers. They are the experts. They want to demonstrate their expertise every opportunity they get. They want to impress the customer with how much they know. They shoot themselves in the foot over and over again. They're also afraid. They're afraid the customer will think they don't have all the answers. They're afraid to ask questions because they may not get the answers they want. They're afraid of silence. When they ask a question, they must wait for the answer, and they cannot stand this silent waiting. They cover this fear by talking too much, many times talking themselves out of the sale.

Most salespeople also make assumptions. They assume the reason for the purchase is the most obvious, logical one. It seldom is. Asking why uncovers the real reasons behind the purchase. Almost every time the real reason is an emotional one.

Logic has its place. It's used to justify the emotion. Asking why uncovers the emotion. This gives the salesperson a tremendous advantage over the competition. When you know the real, emotional reasons your customers purchase a piece of equipment, you hold the keys to the sale. Your competition has to be at least as smart as you are to find that same emotion.

DEVELOP A DYNAMITE SALES PRESENTATION

If you want to develop a red-hot sales presentation, follow the procedure below. It takes time and effort, but it will pay off handsomely in bigger and better sales.

1. Write out your full sales presentation (60 minutes or more).
2. Tear it up!
3. Rewrite it to fill 30 minutes.
4. Tear it up!
5. Rewrite it to fill 15 minutes.
6. Tear it up!
7. Rewrite it to fill 5 minutes. (You have now come to the real meat of your presentation.)
8. Read it aloud. Make any adjustments you think are necessary.
9. Rewrite it using only questions.

You now have a dynamite sales presentation that will command attention and get you sales faster.

WHY ASK WHY AFTER THE SALE HAS BEEN MADE?

Does your supplier send out customer satisfaction surveys after sales have been made? I have had an opportunity to examine a number of customers labeled "likely to defect" because of

P.A.S.S. C.A.L.F.

the nature of the complaints on these surveys. More than fifty percent of these customers would not have been "likely to defect" if the salesperson had asked why after the sale. This is a simple procedure. All it requires is for you to ask the customer, "Why did you just buy that?"

Most salespeople refuse to do this. Why? Because it sounds stupid. Why should you have to ask someone why they bought? Is it not obvious? No – it's not that obvious. No matter how well you explain or sell your product, there is always the chance that the customer misunderstood something. There is always the chance that the customer has developed some unrealistic expectations. You discover this only when the customer satisfaction survey gets back to you. If only you could have known ahead of time! Well, this is the way to find out. It allows you to uncover unrealistic expectations, and it gives you the opportunity to correct them before the goods are delivered.

"Why did you just buy that?"
"What do you expect it to do for you?"
"What else do you think it will do?"

THE GOOD SALESPERSON

The good salesperson knows it's better to ask than to tell.
He knows it's better to be sold than to sell.
So he asks questions.
He asks why.
And the customer sells to him.
And guess what?
Both are happy.

These questions eliminate the unfulfilled expectations that are the cause of much customer dissatisfaction. They guarantee happy customers and eliminate complaints after the sale. Many salespeople resist this approach because they are happy to have the sale and want to get out before the customer changes his mind. This is shortsighted and bad business. Asking why after the sale reinforces the emotional needs of the customer and provides the logic necessary to justify the purchase.

WHAT IS THE BEHAVIOR?

Always ask why the farmer wants to buy anything. Develop "why" questions. Ask why the customer buys twice in the selling process: once before the sale is made and once after the sale has been made. If you do, and keep it up, you allow farmers to sell you—and themselves—on buying your equipment.

THE KEY TO ASKING QUESTIONS

Salespeople are afraid to ask questions. "Nobody likes questions fired at them," they tell me. They're right. If you fire one question after another at a prospect, you risk turning him off. The key to asking many, many questions is to keep your tone conversational. Turn statements into questions. Use a statement to begin another question.

"Mr. Farmer, you say the turning radius on this tractor could save you time. Is that important to you?"

Remember, when you ask questions instead of making statements, customers feel you have their interests at heart. They think you're a good salesperson, and they think you care about what they really want. Isn't that the way you would like your customers to feel about you?

P.A.S.S. C.A.L.F.

WHY IS "ASK WHY" A SUCCESSFUL SELLING BEHAVIOR?

This behavior requires action on your part. You cannot passively make it happen. It requires several other actions as well. It requires that you spend time learning your customer's business so well that you can know when his answers are the correct ones. It requires a conscious effort to change the way you think about selling. It requires you to bite your tongue when you are about to tell a customer something. It requires you to consciously think through your sales presentations and convert them to this higher level of selling.

This behavior will take you out of your comfort zone. It may feel strange and uncomfortable when you try it for the first time. Take heart! If you practice it and perfect it, it will raise your level of selling way above that of your competition. This is the one behavior that can magically do that. But it requires work.

When this behavior becomes a habit, you will have raised your selling abilities several notches. When it becomes the way you do business and the way you sell, your competitors will wonder what hit them.

IF YOU ARE NOT AN AGRICULTURAL SALESPERSON

Of all the behaviors, this one has the most universal application. It is not new or unique. Superstar salespeople who have discovered its magic are using it every day.

Don't think this behavior works only on consumers. Try it in the boardrooms of your corporate clients. In fact, the higher up in an organization you go, the less logic there is in buying decisions. They won't tell you this, of course. Logic is a powerful justifier, but emotions make the sale. Every time.

If you are not currently practicing this behavior in your selling, you may be missing out on some spectacular sales. Once perfected, it makes selling much more fun. It can also shorten your selling cycle dramatically.

BEHAVIOR #7
LEARN ABOUT COMPETITION

HOW MUCH DO YOU KNOW ABOUT YOUR COMPETITION?

Most companies do a tremendous job of providing product knowledge on their equipment. I'm constantly impressed by how much technical knowledge salespeople have on their own equipment. This is a tribute to technical trainers.

I'm also amazed at how little salespeople know about their competition. They can recite all the features, advantages and benefits of their own piece. They know all its uses and applications. But when it comes to comparing those features, advantages and benefits to another brand, many are woefully ignorant.

This places them at a disadvantage when faced with competition. As a result, they avoid talking about those comparisons and avoid the competitive owner. Competition, however, is not just another color. Sometimes the competition is the same color and just down the road. In this chapter, competition means not only other colors but also other in-line competitors. Therefore,

P.A.S.S. C.A.L.F.

learning about them is as important as learning about the other colors. Knowledge is power.

> **THE MORE YOU KNOW ABOUT YOUR COMPETITION, THE LESS YOU NEED TO KNOW.**

WHY SHOULD YOU LEARN ABOUT YOUR COMPETITION?

As self-evident as this question is, it may be a good idea to state some reasons.

1. You're a professional.
2. Learning is part of your job.
3. It's part of your business.
4. You can whip 'em if you know 'em.

To make this behavior work, you must want to know more about your competition.

HOW DO YOU LEARN ABOUT YOUR COMPETITION?

There are several ways you can learn about your competition.

1. One of the easiest is to use the resources provided by your supplier. All kinds of videos, literature, sales manuals and audio tapes are available.
2. Pick up literature on competitor models and study it. Compare what it says about their equipment with what you

know about yours. How is yours better? Which features can you make a strong case for? If you were a buyer, why would you buy it?
3. Ask your customers. Ask those who have purchased from your competitors. Ask them why they bought, what benefits they got and if they are happy with their purchase. Ask those who did not purchase from the competition. Ask them why they did not. If what they tell you can help you in another sale, ask if you can quote them.
4. Ask your competitors' customers. Ask them why they purchased. Ask what the biggest selling point was that got them to commit. Ask, "If you had to do it all over again, would you still purchase this? Why?" Tell them about some features your product has and ask them to help you compare.
5. Read your trade magazines. Read about your competitors. Study their ads. What do you like about them? What do you not like? Track their inventory through the ads. Are the same pieces showing up, or are they bringing in more than you? Read about other equipment. Make sure you know enough about it to be able to talk about it. If you don't understand something, ask your colleagues or sales manager.
6. Drive onto your competitors' lots. Look around. Do they have a lot of astronaut equipment? Do they have good equipment? Come back each month and see how much of their equipment is moving and how much is staying behind. If you know your business well, you should be able to tell a great deal about your competition just by driving onto their lots once a month.
7. Discuss with your colleagues and sales manager. Teach them as well as learn from them. Don't ignore the tech-

P.A.S.S. C.A.L.F.

nician who may have worked on other color equipment. Talk to the parts person who sells parts to competitive owners. The amounts and types of parts being sold can teach you something.

Use these resources and any others you can think of to improve your knowledge of your competition. Incidentally, it drives your competition crazy when their customers come to you for advice on their equipment because it shows you know more about it than they do.

WHAT COLOR IS YOUR PROFILE?

All salespeople use jargon specific to their industry. Agricultural salespeople are no exception. For those readers who are not familiar with some of the terms I use elsewhere in this book, here is a brief lesson in the language of agricultural sales.

Colors: Until recently, farmers tended to show a great deal of loyalty to one particular brand of agricultural equipment. Brands are identified by the predominant color of the equipment—for example, John Deere is green, Case IH is red, and so on. Hence the use of the term "colors" to describe competitors.

Profiles—Data gathered on a farmer, including the number and types of equipment he owns and other relevant information. Profiles are designed to help the dealership better service its customers.

Jockeys—Wholesale dealers of used farm equipment.

HOW DIFFERENT ARE YOU FROM YOUR COMPETITORS?

I conduct an interesting exercise in dealerships. I present a scenario and then ask a question.

Here is the scenario: I am a farmer in the market for a new piece of equipment. You have several competitors.

Here is the question: "Why should I buy from you and not your competitors?"

I have asked this question in over 100 dealerships and the answers are pretty standard—good parts availability, good technicians, long hours, great service, honest people, location, "we care," and so on.

I then ask another question: "If I were to ask your competitors the same question, how would they answer?"

They generally say they would answer the same. That's true. These are all the same things your competitors say about themselves. This would present a problem for me if I really were a customer. I would have to ask myself, "Which one of you is lying?" Or, "Are you all lying?" Or worse, "Are you all telling me the truth?" Because if you were all telling the truth, then there would be no special reason for me to come to you.

You should ask yourself why a customer should buy from you and not your competition. If you cannot answer, you should take a long hard look at yourself—and the competition. If you don't know what sets you apart, what makes you different from your competition, you don't know enough about your competition. How can you set yourself apart when you don't know what to set yourself apart from? You should know all your competitors. It's part of your job.

P.A.S.S. C.A.L.F.

ELIMINATING COMPETITION

If you know you have competition (and you do), and you know your competition can disrupt your sale (and they can), then it makes sense to eliminate them as quickly as possible, right? How can you? The answer lies in comparison. But to be able to make the comparison, you must already know a great deal about the competition. This strategy works well when you have several competitors for the same piece of equipment and your prospect is—or could be—aware of them.

Let's say you have ten possible competitors on a piece of equipment. Select the best two competitors. Then follow the steps below in exact order.

Step 1
"Mr. Prospect, as you are probably aware, you could choose from about ten different tractors to do the job you want done. Some will do it better than others. If I were you and if this was my money I was spending, I would select one of these three."

What have you just done? You have eliminated all the others!

Step 2
"Let me tell you why. This one..."

Explain all the useful features of the BETTER competitor.

Step 3
"Now let me explain why I would suggest this one..."

Explain all the features of the LESSER competitor.

What have you just done? You've eliminated the lesser competitor! Why? Because your prospect is already comparing the better competitor to the lesser competitor and has decided the better competitor is the better buy.

Step 4
"Let me show you why I would choose this one..."

Demonstrate the value of your product. Show all the advantages of owning your product.

What have you just done? You've eliminated all your competitors!
And you never once said anything bad about them!
Will this work for you every time? Nothing works every time in sales. But this can help you eliminate competition most of the time.
What if the competition is in-line? The same principle applies. Look at Step 1 and substitute:

"Mr. Prospect, as you are probably aware, you could choose from about ten different dealers to get the tractor you want. Some are better than others. If I were you and if this was my money I was spending, I would select one of these three."

Can you see how this technique can work in many competitor situations?

P.A.S.S. C.A.L.F.

WHAT IS THE BEHAVIOR?

Spend two hours each week learning about your competition. Don't fall into the trap of telling yourself you are learning all the time anyway. Set aside "learning" time. Make an appointment with yourself. Do whatever it takes to do it—but do it.

WHY IS LEARNING ABOUT YOUR COMPETITION A SUCCESSFUL SELLING BEHAVIOR?

This behavior requires action on your part. You cannot passively make it happen. It requires you to want to be more knowledgeable about your business. It requires you to want to be a professional in your field, to be the best in the industry. Learning can be both active and passive. This behavior is not about passive learning. It's about taking control of what you learn and when you learn it. It's about making the effort when others would rather coast.

This behavior will help set you apart from your peers. It will also help make you a resource not only in your dealership but also to your customers—and your competitors' customers. Once the two hours a week become a habit, it becomes the way you do business.

IF YOU ARE NOT AN AGRICULTURAL SALESPERSON

Once again, take out the agricultural terms and this becomes a behavior every salesperson should pursue. It makes a lot of sense, especially if you sell highly competitive products or services (who doesn't?). After all, your customers expect it from you. One of the reasons customers come to salespeople is that they assume the salesperson has this knowledge. Would you like to be the one to tell your customers you don't?

BEHAVIOR #8

FIRST-TIME TRADE

DOING THE IMPOSSIBLE

This is a simple behavior. It means you should attempt to close every deal the first time, every time. It means you should ask for the order the first time, every time.

Whoa! I can already hear the experienced salespeople saying this is not possible. How can you push a farmer into making up his mind the first time you discuss his needs? This is not the way it works. He needs time to think about it, discuss it with his wife or hired hand, shop around, consult your competitors, or just sleep on it. If you try to push him at this stage, you could lose the sale altogether.

They're right, of course. This is not the way it is usually done.

Informal studies conducted by my good friend Bob Hilleque and myself reveal that it usually takes 4 - 5 calls on a customer before an agricultural sale is made. Lawn and grounds care equipment sales require 1 - 2 customer interactions before comple-

P.A.S.S. C.A.L.F.

tion. It takes a lot of time on callbacks and discussions to complete many agricultural sales. So where do I come from when I say close the deal the first time? Conventional wisdom says it can't be done.

Remember, however, I didn't say you would always succeed in getting the sale. All I said was that you should ask for the order the first time. Will you always get the order? Obviously not. But what happens when you do? It will probably be the most profitable sale you make that day. If you close only 10% of these deals, you're way ahead of the game. Besides, what is the worst thing that can happen? The farmer says no. Big deal! You didn't have the sale at that point anyway. At least you gave yourself a chance of getting it and getting it quickly. This is like the lottery example on page 57. You can't win if you don't buy a ticket. Asking for the order is like buying a ticket. Your odds are fifty-fifty. He will either buy or he won't. If you don't ask, he will definitely not buy at that time.

WHAT HAPPENS IF YOU DON'T CLOSE THE FIRST TIME?

When you look at what can happen if you don't close the deal the first time, you can see why you should at least make the attempt. What can happen?

1. You give your competitors an opportunity to bid on your deal.
2. You decrease your chances of ever getting the deal.
3. Once the farmer starts shopping around, he becomes a more difficult negotiator.
4. The price usually drops after each customer contact.

Roland Barr, an experienced and very savvy salesperson in Ohio, gave the most compelling reason for making the attempt to close the first time. He said, "If you don't trade at the first meeting, it usually gets harder later."

HOW?

The "how" of this behavior is actually very easy. Simply ask for the order! "Mr. Prospect, what would it take for us to do business today?"

But as easy as this may appear, many salespeople have difficulty doing it. Even after they have bought into the concept, they hesitate to put it into practice. Even after they have determined it's what they should do and have every intention of doing it, they wilt when face to face with the customer. Why? They may not believe in the magic of asking for the order. They may harbor secret doubts about its effectiveness. They may suffer from Sales Call Reluctance® (see sidebar, page 92). Or perhaps they fear the objections that may follow.

I can already hear these salespeople shouting, "Not 'may' follow, 'will' follow!" Okay, let's look at these objections.

DEALING WITH PRICE OBJECTIONS

Since most salespeople tell me their biggest objection is price, let's see if we can find a way to deal with it. First, let's agree on some things:

1. Whenever you get an objection, you go on the defensive. However you answer it, you sound as if you're defending a position once the objection has been voiced.

P.A.S.S. C.A.L.F.

2. It's always better to show value before you talk about price. (If only your customers could understand this!)
3. Most price objections are not objectively valid. If they are, then you are overcharging. (Think about it.)
4. It is always better to prevent an objection from coming up in the first place.
5. Price is hardly ever the real objection, even though it is used the most often.
6. The price objection comes up more often than we like.

If all of the above is true, then it makes sense to pre-empt the objection. You do this by bringing it up first. Why should you? After all, what happens if the farmer doesn't bring up the price objection? Aren't you putting it into his or her mind? Yes. But if, as most salespeople tell me, the objection is going to come up more than 80% of the time anyway, then you have only a 20% chance of being wrong. I can live with those odds.

So how do you pre-empt the objection? How do you bring it up first? Here are some examples.

> *"Mr. Prospect, you could shop around, and you might get a slightly better financial deal, but is that all that matters to you? What else is important to you when you buy something like this?"*

> *"Mr. Prospect, I realize price is important, and I'm sure I can structure a deal to satisfy you. However, besides price, what else is important to you?"*

> *"Mr. Prospect, I know my price is a little higher than my competition – and I make no apology for this. When you're the best and you sell the best, you must charge*

a bit more to make sure you can deliver the quality of service you stake your reputation on. Even so, when we get to the price, I think you'll be pleasantly surprised at how little that difference really is considering what we offer."

Did you notice something important in all the above?

1. You have brought up and answered the price objection on your terms. You have not had to adopt a defensive stand.
2. There is no longer any need for the customer to bring up the price. You already have.
3. You are not apologizing for your price. You are taking pride in it and demonstrating why.
4. You are effectively moving away from price.

If you forgot to pre-empt the objection and it comes up, here is something you can say:

"I'm too expensive? I can see how you might perceive that. But think about it. If I were too expensive relative to my competitors, I wouldn't still be in business. They, and my customers, would have run me out of town a long time ago. The fact that I'm still here—and the most successful dealer in the area—says I'm not too expensive. Most customers are like you. They're no fools when it comes to value. That's why they keep coming back. They don't begrudge me a little extra because they know it buys them the quality service they have come to expect from me.

P.A.S.S. C.A.L.F.

"Do you also demand—and expect—top quality service long after the sale is over? Why?"

WHAT IS THE BEHAVIOR?

Make a conscious effort to ask for the order the first time, every time. Make at least one attempt to close every deal the first time. No exceptions.

ARE YOU A YIELDER™?

Perhaps one reason salespeople hesitate to ask for the order is something called Yielder™ Sales Call Reluctance®. In the years I have spent working with agricultural dealer salespeople, I have found many unresolved cases of Yielder™ Sales Call Reluctance®. In fact, more than 70% of those we tested showed elevated Yielder™ scores.

What is it? It's one of the 12 types of Sales Call Reluctance®. It's a learned fear that prevents good salespeople from acting assertively with customers. It impedes prospecting (an assertive act), closing (another assertive act), telephoning a customer or prospect, asking for money, asking for a commitment, making definite appointments and calling on competitive owners. It forces proud salespeople to constantly apologize for selling good equipment, and it makes them vulnerable to price issues.

Fortunately, many agricultural dealer salespeople have found a simple, easy solution to this scourge. Nearly 4000 dealer personnel around the country have already attended the remarkable Fear-Free Prospecting and Self-Promotion Workshops® developed by Behavioral Sciences Research Press. Many have learned to manage Yielder™ and other types of call reluctant behavior. If you feel this may be what prevents you from performing at your best, diagnosing and correcting it should be one of your first steps toward recovery.

WHY IS FIRST-TIME TRADE A SUCCESSFUL SELLING BEHAVIOR?

This behavior requires action on your part. You cannot passively make it happen. It requires that you examine the way you make sales and develop ways to ask the right closing questions at the appropriate times. It requires that you practice your close with colleagues or your sales manager. It requires you to assert yourself at the time of the close. If you have difficulty with assertion, it requires you to learn to manage your fears. It requires you to learn about timing.

You should practice this behavior until it becomes a habit. When it does, you will increase your sales more than you can imagine. It will make selling more fun and profitable. When it becomes a habit, it becomes the way you do business. First-time trade will leave your competitors wondering. You will have cut them out of deals they will hear about only after it is too late.

IF YOU ARE NOT AN AGRICULTURAL SALESPERSON

In any sales situation it makes sense to attempt a close at the first meeting. Even when it does not appear to be appropriate, no harm can come from asking for the order. In some cases, getting the order may not be possible at the first meeting. Complex sales, for example, preclude this. Even in these cases, however, it is appropriate to close on something, to ask for some commitment.

While this behavior may at first appear to apply to other sales positions and not yours, look again. You may be missing out on some good sales. Consider something else: Perhaps the prospect wants you to close immediately. Perhaps he or she wants

P.A.S.S. C.A.L.F.

to save time, too. Give prospects the opportunity to say yes or no. They have a right to this opportunity.

As for price, this is not just an agricultural issue. In every industry I have consulted, price is almost always the number-one objection. If it's not at the top of the list, it's usually not far from the top. This is a universal problem and one that makes very little sense. If what you are selling will not benefit the person you are selling to, then the price does not matter. It will be too expensive no matter how low you are prepared to drop your price. You should not even be talking.

If, on the other hand, your product or service will help fulfill an important need for your prospect, again price does not matter. Of course, your price has to be within reason, but it does not matter that you are more expensive or less expensive than your competition.

If price were the only factor in a sale, then only the cheapest would sell and the rest of us would be working for that company. John Deere salespeople would be working for their less expensive competitors. Designer clothes would become extinct. Hyundai would dominate the automobile industry. (There goes your Lexus!) And the wildly exorbitant (but deliciously luxurious) Peninsula Hotel in Hong Kong would become a Super 8 Motel.

Study after study tell us that price is not the most important factor in purchasing decisions. Over and over salespeople ignore those studies in the field while paying lip service to them in the classroom. The complete salesperson understands the value of price in a sales transaction and uses it to his or her advantage.

CONCLUSION

There you have it—P.A.S.S. C.A.L.F.—eight simple steps to sales success in agriculture or any other industry. This is not rocket science. These are behaviors each of us can, and should, practice on a consistent daily basis.

Develop these behaviors one at a time. Spend at least two weeks developing one before going on to the next. Remember, however, you want to develop the behavior to become a habit. Don't abandon one to go to the next. As you master each one, continue developing it into a habit as you begin on the next. Will it be easy? Probably not. But then nothing worth having comes that easily. With just a little effort and willingness on your part, it is definitely possible.

Develop the behaviors. Make them habits. They will make you money.

Simple? It always was.

ABOUT THE AUTHOR

Frank Lee has taught thousands of people the value of "Beyond FAB" selling.

Other salespeople, and such notable authors as George W. Dudley, have hailed Frank as a sales superstar, a master salesperson, and a great sales manager. However, Frank will tell you his skills are no more than average. Certainly, he knows and understands basic selling strategies. He will admit to many years of diverse selling experiences. He will even confess to having been extremely successful as a salesperson and sales manager.

What he attributes his selling successes to is something of a surprise. It is not, as he says, due to his "dazzling charm, wit and intelligence." Rather, it is something far more basic and achievable. Frank credits his successful selling career to practicing successful behaviors on a consistent daily basis. These successful behaviors have helped him create and run businesses in seven countries. They have catapulted him from a start-up operation to one of the largest sellers of the Fear-Free Prospecting and Self-Promotion Workshop® in the world—in less than three years.

These behaviors are not only simple, they are achievable by any dedicated salesperson.

This book is a combination of the following:
1. Frank's years of experience using successful behaviors; plus
2. His experience traveling with many agricultural sales people; and
3. Numerous sales workshops he has taught in eleven countries.

Today Frank Lee continues to sell in huge quantities and to teach salespeople in many industries how to make "successful behaviors" work for them, too.

He lives in Flower Mound, Texas but regularly travels all over the United States and overseas. He is married and has two children and two dogs, all of whom he is extremely proud of.

POWERFUL SALES PROGRAMS FROM SALES ACADEMY, INC.

THE SALES CALL RELUCTANCE® PROGRAM
When well-trained, educated salespeople fail to deliver on the promise sales managers and recruiters forecast for them, it is usually not due to a lack of skill or knowledge. Most times it is due to Sales Call Reluctance® – the emotional short circuit that prevents capable would-be sales superstars from ever reaching their true potential. Even experienced salespeople can suffer the debilitating effects of unresolved cases of Sales Call Reluctance® that put their careers on permanent hold.

SPQ*GOLD™: THE SALES CALL RELUCTANCE® SCALE
Laser-sharp and amazingly accurate, this test identifies the reasons why salespeople do not prospect and provides sales managers and trainers with the tools necessary to better select, develop and manage their salespeople.

THE FEAR-FREE PROSPECTING & SELF-PROMOTION WORKSHOP®
This unique one-day workshop helps salespeople discover and understand how one or more of the 12 types of Sales Call Reluctance® can sabotage their careers. They learn to manage their fears so these fears no longer prevent them from making the calls necessary to sustain successful sales careers. All participants receive a Personal Prescription Profile™ that lays out practical ways to beat the Sales Call Reluctance® demon once and for all. Managers learn how to follow up with salespeople to ensure the effectiveness of the workshop.

MANAGEMENT TRAINING WORKSHOP
Sales managers and trainers have called this "the most powerful workshop on earth" for good reason. Unlike any other workshop anywhere, this three-day experience prepares trainers to facilitate the Fear-Free Prospecting & Self-Promotion Workshop® and sales managers to manage salespeople with Sales Call Reluctance®.

SALES CALL RELUCTANCE® STUDIES AND RESEARCH
Sales Academy and Behavioral Sciences Research Press undertake specialized studies and research projects with selected client companies to determine the optimum use of the Sales Call Reluctance® program.

TO FIND OUT MORE ABOUT THE SALES CALL RELUCTANCE® PROGRAM, OR TO PURCHASE ADDITIONAL COPIES OF THIS BOOK, CALL:

(800) 898-3743

POWERFUL SALES PROGRAMS FROM SALES ACADEMY, INC.

SALES TRAINING PROGRAMS

Sales Academy takes a different approach to sales training. Following the principle of the Sales Call Reluctance® program – that most salespeople do not fail because of a lack of skill or ability – Sales Academy looks at developing successful behaviors for salespeople. These are behaviors that propel salespeople to new levels of production.

THE BEHAVIORAL SALES WORKSHOP

In preparation for this workshop, Sales Academy studies the behaviors of a client company's salespeople to identify good and bad selling habits. It then customizes a workshop that teaches salespeople the successful selling behaviors most appropriate for their sales situations. Salespeople learn the value of these behaviors. They also learn practical ways to implement them on a daily basis. Sales managers learn how to manage these behaviors so their salespeople can continue to perform consistently at a higher level.

THE SUCCESSFUL DEALER BEHAVIORS PROGRAM

Specially developed for the Agricultural Dealership, this is a four- to five-month exercise in developing managers and employees using behavioral models. Everyone in the dealership participates. Managers learn how to develop and implement non-negotiable and successful behaviors in all areas of their dealership. They learn the two essentials of effective management – getting people to do what they should, and developing them to become better. This highly acclaimed program includes the Sales Call Reluctance program, the Behavioral Sales Workshop and the Most Valued Customer/Teamwork workshop.

MOST VALUED CUSTOMER/TEAMWORK WORKSHOP

Practical, hands-on, totally-customized for each dealership. This workshop creates dealership teams to close the gap between what their top customers spend with them and what they spend in total. It also shows how to systematically acquire additional Most Valued Customers. No theory! Real customers and prospects are chosen. Teams create Strategic Sales Plans in the workshop to make this a reality.

TO FIND OUT MORE ABOUT THESE SALES TRAINING PROGRAMS, OR TO PURCHASE ADDITIONAL COPIES OF THIS BOOK, CALL:

(800) 898-3743